D1200981

PIONEER
ARTS AND CRAFTS

by

EDWIN C. GUILLET

University of Toronto Press

First published by the
Ontario Publishing Co., Limited, 1940

This edition
© University of Toronto Press 1968
Toronto and Buffalo
Reprinted 1971
Reprinted January 1973, November 1973

Printed in the United States of America

ISBN 0-8020-6081-1
LC 72-415879

To

MARGUERITE GUILLET BROOKS

Designer, Thread Workers Guild of America

PREFACE

This volume was first published in 1940 as one of the *Early Life in Upper Canada* series for students of Canadian social history. The development since that time of "pioneer villages" with arts and crafts in action has greatly increased and widened interest in the subject.

Necessity, it is said, is the mother of invention, and our early settlers were ingenious and skilful in providing for their needs in their spare time. The more enterprising engaged in domestic manufactures, and were quick to adopt farm processes brought to their attention at fairs and exhibitions, while their women were equally interested in improved housekeeping and recipes for food and drink.

Prominent among authentic source materials for this book is *The Female Emigrant's Guide* (1854) by that cultured and intrepid pioneer, Catharine Traill; but her brother Sam Strickland and numerous other contemporaries have also been drawn upon in describing ways and means of doing things at a time when perseverance amid discouragement was the rule.

This edition of *Pioneer Arts and Crafts* has a few small additions and a number of new illustrations.

Edwin C. Guillet

January 1, 1968

CONTENTS

ILLUSTRATIONS

PIONEER ARTS AND CRAFTS

(1) THE AGE OF WOOD

THE Canadian pioneer lived in an age of wood. Many complicated machines and contrivances consisted entirely of wood, though wood may form no part of the machines which do their work in the industries of today. Similarly most of the equipment of the pioneer home was hand-made and of wood. When a door took the place of a blanket or some makeshift covering of the entrance to the log house, its hinges, lock, and latch were often of wood. In describing the experience of a settler in Ops Township in raising his log house in the wilderness, Thomas Conant wrote:

To form a door he split some thin slabs from a straight-grained cedar and pinned them with wooden pins to cross slats. The most ingenious parts of the construction, however, were the hinges. Iron hinges he had not and could not get. With the auger he bored a hole through the end of a square piece of wood, and sharpening the other end with his axe, he then bored a hole into one of the logs of the house, constituting in part a door-jamb, and drove the piece of wood into this hole. This formed the top part of the hinge, and the bottom part was fashioned in exactly the same way. Now to the door in like manner he fastened two pegs of wood with holes bored through their ends. Placing the ends of the hinges above one another, they presented the four ends with holes leading through them, the one above the other. Next he made a long pin with his handy jack-knife, leaving a run at one end of it and making it long enough to reach from the top to the lower hinge. Through the holes at the end of the hinge this long pin was placed, and thus the door was hung.

Wooden pumps are still in common use in Ontario, though many farmers have now iron pumps with which water may be forced to various buildings. James B. Scott of Oakland Township, Brant County, who had been making wooden pumps since 1883, used tamarack logs and bored them with a two-

John Ross Robertson Collection

KITCHEN UTENSILS, 1813
Most of them were hand-made in the early period.

inch pod auger. The head of the pump was made of quarter-cut pine. A gasoline engine provided power for the boring, but until 1916 Mr. Scott used a horse to turn the wheel, and some of his early competitors even bored by hand. When he entered the trade there were numerous pump-makers in the district, but at the time of his death in 1941 Mr. Scott was the only one left within a radius of at least fifty miles.

(2) CUTTING BOARDS

Many a pioneer house was floored—if at all—with half logs roughly formed by adze or broad-axe, for smooth boards were available only at sawmills. Some settlers contrived to make boards sufficiently even for partitions by splitting logs known to break with a fairly smooth surface. At times a whip-saw was used, but it was laborious work to saw from end to end of a log, one man standing in a pit with the sawdust falling continually in his eyes as he attempted to keep up with the man above him. Samuel Strickland developed improved methods of *slabbing*, or splitting planks from pine logs. At one end of an eight-foot log he marked off planks four inches wide, and then split them out with wedges, starting from the outside

Conservation Authorities, Ontario
COLEBROOK PUMP FACTORY, NEAR NAPANEE

rather than from the centre. Later he improved his planks by the following process:

I cut a square notch on the top of two large logs laid nearly the length of my plank apart from each other; I then placed one edge of the plank in the notches, which I wedged firmly. By this method, after lining the upper edge, I was enabled to hew the surface of the plank with my broad axe and reduce it to the proper thickness. As soon as the surface was smoothed I struck a straight line on each side of the plank, which I dressed with my axe, thereby forming a square straight edge easily jointed with the plane. In this manner did I prepare my flooring and partitions, which for a time answered a very good purpose, as roof-battens for shingle on narrow cedar boards are easily split. Those I made were six feet long, and varied from four to eight inches in width and an inch thick. A board of this length reaches across three rafters. (The white cedar splits very freely.) I boarded both ends of my house with planks made in this manner.

Sawmills were at first as scarce as grist-mills. By 1792, however, there were thirteen between the mouth of the Trent River and Long Point, though all of them were in the Niagara district, which at that time contained almost all of the settlers in central Upper Canada. As the years passed, the sawmill was often the nucleus of settlements. Thomas Need, whose sawmill was the beginning of the village of Bobcaygeon, says that 'the erection of a sawmill is always the first marked event in the formation of a settlement in the Bush. . . . This induces many to come into the neighbourhood, from the facility it offers for building.' So the old methods of making planks gave way to machinery, and when a farmer took logs to the mill he usually received back half the boards, the other half being retained by the miller as payment.

(3) SPLITTING SHINGLES

Before the use of shingles, hollowed basswood logs or strips of bark commonly formed the roof of the pioneer's log house. The first shingles were about three feet long, and were split with considerable difficulty from blocks of cedar or pine, or occasionally from ash or oak. The frow or the axe was used for the purpose, and the shingles were heavy, usually uneven, and not shaved, serving as sheeting as well as shingles.

Pole rafters of split cedar, four inches wide, were laid a foot or so apart lengthwise across the rafters, and to these the shingles were fastened by wooden pins, each row overlapping the previous one. Barn shingles were often four feet long and placed on round rafters in constructing the roofs of the large cedar-log barns which may still be seen in various parts of Ontario. When eighteen-inch shingles came to be used, they were often nailed to strips of lath laid a few inches apart across the rafters, in place of sheeting, for this allowed circulation of air and kept the shingles dry underneath, delaying the process of rotting.

Experienced settlers quickly learned to pick sections of trees which would split fairly, for the appearance of outside knots, of the grain, and even of the limbs aided in the choice. The most suitable blocks would snap apart easily when split, without deep curves which would make holes in the shingles. About four shingles to an inch was usual, and a good splitter could keep two men busy shaving or dressing, which was best done with a draw-knife. Shingles were commonly packed in bundles of one thousand in a packing-frame.

Many of the most enterprising farmers constructed a shaving-horse, which was a combination of work-bench and vice. It was usually higher at one end than the other, and the worker sat astride the lower and narrower end, while through an opening in the bench in front of him projected the head of a clutch swinging on a pin which passed from one side to the other through the body of the bench or horse. Upon a transverse piece of wood on the lower end of the clutch, the operator placed his feet, and by pressing it from him he brought the head of the clutch down upon any object placed under it, and so held it firmly upon the horse. A draw-knife with handles at both ends was used to shave or cut the wood held in the shaving-horse, and it was particularly valuable in shaving and dressing shingles after they were split. Laths were similarly prepared.

(4) MAKING FURNITURE

For the first few years the furnishings of a log shanty were usually no more elaborate than the house itself. A traveller observed that they consisted of 'a bedstead, roughly hewn out with a felling-axe, the sides, posts, and ends held together in screeching trepidation by strips of basswood bark; a bed of fine field-feathers; a table that might be taken for a victualler's chopping-block; four or five benches of the same rude mechanism; and the indispensable apparatus for cooking and eating.'

Early bedsteads were often built into a corner by fitting small six-foot poles into holes bored in the log wall. The outer ends of these poles fitted into a crosspiece, one end of which was inserted in the wall, and an upright used to support the other. The poles were small enough to give considerable spring to the bed. Samuel Strickland describes as 'the only article of luxury in my possession', a bed made of ironwood poles and plaited elm bark. A more advanced type of bedstead was the *catamount*, made of four posts and four poles. The sides were longer than the ends, and a mattress of woven elm bark or of hemlock boughs was later replaced by a straw tick. Some settlers made a rope bed consisting of cords twisted very tight; when the ropes sagged, a special type of wrench was used to tighten them. Another type of 'spring' consisted of basswood bark woven across the bedstead.

The first mattresses were made of materials ready at hand. Spruce boughs formed the most primitive, while dried beech leaves, moss, and straw were used when the bed-tick was filled. Most early mattresses had a large opening in the upper centre to enable easy adjustment of the stuffing. A feather tick—

considered a great luxury—was sometimes placed on top if
the farmer had poultry. The more primitive beds were none
too comfortable. 'The old cord bed-frame,' says an early settler, 'was a veritable trough, and the only thing that made the
squeaking old thing endurable was a plethoric straw tick.'
Buffalo or bear skins sometimes provided the only bedclothes.

A child's trundle bed was a sort of bunk on wheels, which
could be pushed out of the way under the parents' bed. Sap-
trough cradles were commonly used for infants, but there were

Niagara Historical Society

HAND-MADE CRADLE
The handles were for rocking, and the knobs for securing the covers.

other types woven of peeled hickory bark, withes, and rushes.
Rope handles were frequently found on children's beds, and
some cradles had elaborate hoods, handles at the foot for
rocking, and knobs to which to tie covers. A curious rocking-
chair cradle was constructed by William Hunter of Cavan
Township so that his wife might sew and rock her baby at the
same time. It was made of basswood, the seat of the chair and
the base of the cradle being in one piece fifty-one inches long
and two inches thick and supported by a rocker at each end.
It was still in use over a century later.

Sections of tree trunks provided stools until chairs or benches could be made. At first wooden pins held logs and slabs together. Sometimes the bench was convertible into a bunk, and had also a hinged lid which enabled the storage of the bedding. As the years passed, the more handy settlers built chairs for the home. An early type was strongly made of pine, with square legs tapering towards the floor. Frequently the seats were of split elm or willow bark. Men skilled in the work gradually developed the trade of chairmaker to the community.

Mrs. Catharine Traill wisely observed that elaborate and showy furniture—even if there were money to buy it—was out of place in a log house. She had in her home a barrel-chair, and gives the following details of its construction:

The first four or five staves of a good, sound, clean flour barrel are to be sawn off level, within two feet of the ground—or higher if you think that will be too low for the seat: this is for the front; leave two staves on either side a few inches higher for the elbows; the staves that remain are left to form the hollow back; auger holes are next made all round on a level with the seat, in all the staves; through these holes ropes are passed and interlaced so as to form a secure seat: a bit of thin board may then be nailed flat on the rough edge of the elbow staves, and a coarse covering of linen or sacking tacked on over the back and arms; this is stuffed with cotton-wool, soft hay, or sheep's wool, and then a chintz cover over the whole and well-fitting cushion for the seat completes the chair.

Sofas or couches of pine,* stuffed with wool, were also made by handy settlers during their spare time in winter. The first rough table of boards often gave way to fine drop-leaf tables of walnut or cherry, while hardwood pegs driven into the wall or ceiling, to hold various implements and stores, were eventually replaced by cupboards and chests of various types. One settler at Erindale had three very handy sons who were skilful at the lathe and had brought a box of tools with them from the Old Land.

In a very short time [we are told] they made all the wooden furniture of his new frame house—sofas, and tables of every kind, from a lady's work table (with roped pillars of black walnut) to the kitchen table; chimney pieces, painted, polished, and varnished; bedsteads, carts, waggons, and wheel-barrows—they were also equally expert at smith's work and shod their own horses.

*Though furniture of walnut, butternut, cherry, and bird's-eye and other types of maple retains its popularity, pine has come to be considered the most characteristic Canadian wood, and arrow-back chairs, tables, and cupboards of pine are in great demand among the increasing number of Canadians who are furnishing parts of their homes with the best native antiques.

(5) MAKING VEHICLES, IMPLEMENTS, AND TOOLS

Most of the earliest settlers made their own wagons, sleighs, and farm implements. Wheels for carts were sometimes entirely of wood, for there was no iron with which to bind them. In place of a plough, a drag, consisting of a crotched stick with wooden teeth, served the purpose. Crude hand-made hoes and forks were usual, and wooden yokes to fit the shoulder. Some pioneer farmers made their own cradles for reaping, forming the snath by bending it around a tree. The flail for threshing was also home-made. As it was made of two pieces, the shock as it pounded the sheaves was more readily absorbed, and the worker was able to hit the wheat with the full length. Sometimes a fan replaced it. This was an instrument made of ash boards in the shape of a semi-circle of two feet radius. A rim of about six inches was bent around it, and there were holes at either side for handles. It looked like a large grain scoop, and when grain was shaken in it, the chaff blew away. If there was no wind for winnowing, two men waved a sheet up and down.

Another type of separator called a riddle—usually a screen woven from horsehair—was made to sift meal. In fact there was hardly a need which could not be satisfactorily—if crudely—met by the handy pioneer. Even a wooden velocipede was built in the village of Omemee. There were no springs in it, and its handles, pedals, hubs, spokes, and wheel rim were of wood, only the tires being of iron. It must have been rough riding on such a vehicle over the mud and gravel roads of the day!

Soon every village had a wagon-maker, and the larger firms made a variety of buggies, 'pleasure wagons', and carriages for town and country use. Men skilled in making agricultural implements, like Hart Massey of Newcastle, gradually developed large manufactories, which served an ever-increasing territory.

A great variety of tubs, firkins, buckets, and churns were home-made before tradesmen in the towns specialized in their manufacture. Utensils, such as ladles, rolling-pins, and potato-mashers, were often made at home. A gourd, dried and hollowed out, provided a satisfactory dipper. The early Loyalists used spoons and dishes roughly cut from basswood, and when peddlers came around with pewter spoons, a few settlers obtained pewter and moulds and imitated them. In some settlements thorn spikes were gathered as a substitute for pins,

Canada Farmer, 1864

STACKING HAY BY HORSE POWER

One of numerous agricultural inventions found to be impractical.

EARLY COOPER SHOP, VANESSA

which sometimes cost half a dollar a paper. Many a settler had a workshop and forge at the end of the barn, and there, with such carpenter's tools as he possessed, he made axe-handles, whiffle-trees, ox-yokes, a wheel-barrow, and various pails, bowls, and utensils. The first washing-machine was not a complicated utensil, for it was merely a 'pounding barrel', not unlike a churn. A large wooden pounder and soft soap removed the dirt from the clothes. There was also the 'battle board', upon which the washer battled with the clothes by using a stick flattened at one end; and there were ingenious and effective wringers constructed by the more skilful.

Sand and lye were frequently used in pioneer homes to clean the floor, benches, and tables. The first brooms were primitive, for cedar or hemlock boughs tied to a handle provided both broom and brush. The splint broom was made from a stick of green hickory, birch, or blue beech, which was splintered fine with a knife, as deep as desired; then the upper end was shaped as a handle. The finished broom was usually about five feet long and four inches in diameter. A home-made corn broom was the next development. Scottish settlers in Scarborough Township imported plants of Scotch broom, though the brooms were, in this instance, designed mainly for use in their favourite winter sport, curling.

Mrs. Traill describes the heroic efforts of a widow and her children as an example of enterprise and perseverance amid misfortune. The father had been killed by a falling tree, leaving six children under fifteen years of age, but in addition to planting wheat, corn, and potatoes and looking after their cattle, they chopped and underbrushed several acres of new land and erected a brush fence around their crops. The boys were handy with tools, and spent their spare time shaping axe-handles and hickory whip-handles, which they sold to the nearest merchant; and they made brooms of ironwood, blue beech, or oak for neighbouring farmers, while the mother knitted socks and made slippers of listing. The family also made maple sugar and syrup, and the boys obtained groceries and clothing in exchange for loads of oak and hemlock bark, which was in demand for tanning. Soon this enterprising family was able to pay further instalments on the farm which their father had bought just before his death.

(1) CLOTHES OF HIDES AND SKINS

THE reminiscences of Loyalist settlers suggest that the use of skins to make clothing was widespread. While this appears to be somewhat exaggerated, there is no doubt that they were frequently used when other clothing was scarce. Buckskin breeches and jackets, skirts and petticoats, gloves and mittens, and coonskin caps and bonnets were sometimes an important part of the family wardrobe. There were deerskin saddles and bed-covers, and buckskin moccasins provided the first footwear. Sometimes the skins were well tanned into leather, and such useful articles as thread and drinking vessels made from them. One settler says that he wore a pair of leather trousers for twelve years and then sold them for $2.50, for they were still in good condition.

(2) TANNING LEATHER

The skins of animals have been tanned since the dawn of civilization. Drying the skins was found of value in delaying decomposition, but some other means of preservation was needed, and the discovery that the bark of most trees was effective for the purpose was probably made by accident. The tannic acid in the bark combines with the gelatine, which is the chief element in hides, to tan them into leather. Oak bark was long considered the best for the purpose.

The American Indian tanned skins after this fashion. He took the ashes from his campfire, put water on them, and soaked the skins in the solution. In a few weeks the hair and bits of flesh came off, leaving the raw hide. By working the hide with his hands and rubbing it with sticks he made it soft and pliable. The brains of certain animals were sometimes rubbed into it for the same purpose. The tanning, in a solution

TAILOR F. R. DAVIS AND STAFF, LITTLE CURRENT, 1890's

JAMES FISHER'S HARNESS SHOP, GORE BAY, 1890's

of hemlock and oak bark, took about three months, and the leather was again worked by hand.

The pioneer settler followed the Indian method with some variations. The hair was removed by soaking the hides in water to soften them, and they were then laid in piles in a heated room for a time, when the hair would become loose. Usually the hair was merely scraped off, but milk of lime was sometimes used as an aid.

The hides were then placed in a tan-pit, alternated with strips of oak, hemlock, or alder bark. The pit was filled with water, and the process of tanning allowed to continue for several months, after which they were rubbed with bear-oil or tallow to make them soft and pliable. Besides the use of leather for harness and boots, leathern aprons were commonly worn by tradesmen at their work, and farmers found many purposes which it served.*

Most settlements had tanneries, which were usually built near waterfalls so that there would be power to grind hemlock bark. In 1804 Philemon Wright began to tan leather on a large scale at Wright's Mills (Hull), obtaining a cylinder from New York to grind the bark. The Lowell Tannery, on North Yonge Street near Newtonbrook, was early prominent, but as a modern tanning establishment it uses many new methods. The chrome or chemical process is now employed for light leathers. The hides are stored in refrigerated warehouses and preserved by salt until used. The salt is then removed by soaking, lime and sodium sulphite loosens the hairs, and the hides are thoroughly scraped and cleaned in a chemical solution. Several baths tan the leather, and oil emulsions give softness and pliability. There are numerous finishing processes of graining, glazing, and colouring. Altogether, the industry is carried on in a manner almost entirely different from that of pioneer days.

(3) MAKING SHOES AND HARNESS

Many of the earliest settlers crudely cut and sewed or pegged the boots for their family. German and Dutch immigrants

*Many settlers from the Old Land were skilled in the tanning and shoemaking processes. James Boswell noted in 1773 that in the Isle of Coll, Scotland, 'Every man can tan. They get oak and birch bark and lime from the mainland. Some have pits, but they commonly use tubs. I saw brogues very well tanned, and every man can make them.' (*Journal of a Tour to the Hebrides with Samuel Johnson, LL.D.*)

sometimes wore wooden clogs upon arrival in Canada, but soon learned that they were unsuitable for conditions here; while the Englishman found that pegs were often used in America because thread rotted so quickly amid the swamps and mud of settlement. Sometimes a man who was particularly skilful at the work would become an itinerant shoemaker,* travelling from house to house to make a year's supply of shoes for the family. He carried his own kit of tools, consisting of a hammer, a knife, a heavy needle, awls for punching holes, heavy nails, thread, and a rough wooden last around which to

From *Pioneer Days*   Creigh Collins

MAKING SHOES

shape the shoe. A small tub of water stood near by to soak and soften the leather which the farmer had prepared for the shoe-maker's use. As he pounded and sewed, a strap around his foot and the shoe held it firmly in position, allowing him full use of his hands. The shoemaker usually boarded with the family.

*At the middle of the century William Matheson, who lived in Stephen Township, Huron County, travelled on foot in that region repairing watches and clocks. Cash was scarce, but settlers paid if and when they could. Sometimes, if logging or harvesting were in progress, he would stop a few days and lend a hand. Later Matheson set up a shop in Lucan, Biddulph Township, adding photography to his other skills.

Harper's Weekly, 1870   John Bolles

THE TRAVELLING TINKER
Itinerant tradesmen were characteristic of the pioneer period.

KETCHUM'S TANNERY, YONGE AND ADELAIDE STREETS, TORONTO, 1834
At the left are the currying shed, bark mills and piles of tan bark;
and at the right Jesse Ketchum's office and home.

James Treasure wrote home from Yarmouth Township, Upper Canada, in 1830, telling of his success as a shoemaker:

> I have a great deal more than I can do now, and they tell me it will come in faster after harvest; but there is no possibility of getting hands. I have 13s. 6d. for making a pair of Wellington boots, the leather being found me. . . .The price for making men's and women's shoes is both alike, 4s. 6d. for light and 3s. 6d. for strong. They find [*i.e.,* provide] their own thread, too. I can now save money very fast, and shall soon be able to buy my own leather, which will be more profitable.

With the growth of settlements the shoemaker lived in the village, and there was also a hatter, a saddler, and probably a harness-maker. Frequently the currier had a separate shop where he brushed, scraped, and rolled the leather, making it soft and pliable. It was then passed on to the saddler, who did his work on a sewing-horse, or to the harness-maker, who cut and stitched it much like the shoemaker. Eventually the currier's work was carried on by the equipment in the tannery, and in modern times large factories have been built, doing by machine the work formerly done by small tradesmen.

CHAPTER 3 SPINNING AND WEAVING INDUSTRIES

(1) SHEARING AND SORTING WOOL

SHEEP were not raised in the pioneer period without great difficulty. Wolves were common in many districts, and winter feed was sometimes little but pine boughs; but if the sheep survived the cold winter they usually had a fine thick fleece. Most of the work connected with the wool industry fell to the lot of the women. When sheep-shearing time came in the late spring the sheep were carefully washed in a nearby creek. Sometimes it was necessary to apply a solution of tobacco leaves to destroy vermin. After a couple of days the sheep were thoroughly dry and the shearing began, hand shears being used. Joseph Pickering wrote as follows concerning a sheep-shearing in early June:

Sheared the sheep today; the Canadians shear the belly and neck, and then tie the sheep's legs and shear along them, or rather any or every way, and are not nice about their appearance when finished. One Canadian sheared near fifty.

Shearing was followed by picking and sorting, when all matted, coarse, and dirty wool was removed from the clean, later to be spun into coarse yarn for rough purposes only. A 'picking bee' was often called, when the neighbours gathered for the work. It was a jolly occasion, for a bee always included gossip, singing, and dancing.

(2) GREASING AND CARDING

To aid in combing, the wool was carefully greased. Lard, oil, or refuse butter was used, being melted and poured or sprinkled over the wool. By stirring and rubbing the wool with the hands, using three pounds of grease to seven or eight of wool, the workers distributed it evenly throughout. Whipping with a rod was another method of working the grease evenly into the wool, which was then rolled up in blankets, ready for carding.

The carding process consisted of combing the wool into rolls by the use of carding-boards, one with the teeth upward and the other down, which were drawn back and forth until all the fibres were brushed straight. The wool was then twisted or rolled into a soft ball or *roving*, ready for spinning. Even in the eighteen-thirties most districts had at least one carding mill, where for 2d. to 3d. a pound the work was satisfactorily done, and farmers received their wool back to spin into yarn.

(3) SPINNING

The wheel used in spinning woollen yarn was a large one, so the operator stood as she worked. The wool was placed on the point of the spindle, and the wheel worked with the right hand, perhaps aided by a small forked stick, while the left held the spindle. As the wheel whirled around, the operator walked three steps back and forward, and the wool twisted and the yarn wound on the spindle. In a day she might walk twenty miles while spinning half a dozen skeins of yarn.

I like the hum of the spinning-wheel amazingly [remarks an early traveller], and have often waited to look at some tidy girl walking backwards and forwards at her task, at each approach sending off another hum as she drives the wheel round once more.

Yarn for knitting was spun only once, but for heavy cloth it was usually spun twice. The reel was used to wind the yarn into skeins, and one of the daughters of the house was often engaged at the reel while her mother worked the spinning-wheel. Each time a knot was wound on, the machine clicked when an upright slat was released and struck a nail. Fourteen knots made a skein, and four skeins of wool was a good day's work at the reel. Sometimes a machine called the swift was employed to unwind the skeins when the yarn was being wound into balls, but the yarn was usually hung up for a time in skeins.

Preparing wool and spinning provided work for the whole family. Handy boys frequently made the loom and reel spools with their jackknives, and they usually aided also in the hard work of washing the yarn and removing every particle of grease.

In some parts of Ontario spinning-wheels are still used, but carding mills where the wool is prepared are now few and inaccessible, and wheels and reels, if they have survived at all, have become curiosities stored away in the loft over the old driving-shed.

(4) DYEING

Dyeing originated in countries of the East, but the Canadian pioneer added many distinctive dyes to those of ancient times.* The process sometimes took place after the weaving, but it was usually more satisfactory to dye the yarn. The common 'sheep's-grey' flannel and fulled cloth worn by the men was made without dyes, by mixing one part of the wool of the black sheep with three of the wool of the white, or other proportions as desired.

Before attempting to dye yarn all grease and oil had to be removed from it by scouring in hot soap-suds. The soap, too, must all be rinsed out in soft water if the colour was to be equally distributed. Iron vessels were employed for most dark colours, but brass or tin for light and delicate tints.

Dyes were usually home-made. White maple bark, boiled and set with alum, gave a good brown-grey. A slate-coloured dye for cottons was produced by soaking in sumach-bark liquor, and then in warm water in which a little green copperas had been dissolved; a weak logwood dye with a little pearl ash gave a bluish tinge, while Brazil wood made it lavender.

Purslain weed and logwood, set with alum, gave a bright blue dye, but indigo plants were usually grown for the purpose. Madder provided red, and waxwood, onion skins, horse-radish leaves, smartweed, and goldenrod flowers were used for yellow. Sumach blossoms, walnut husks, and certain barks like butternut were employed to produce brown dye, the inner bark of the butternut being steeped in cold water for several days, and yarn or cloth soaked in the strained liquor.

Beech bark provided a drab colour, while green was produced by steeping in yellow dye and then in blue, and there were various other combinations according to the desire and skill of the dyer. The lye of wood ashes, with a little copperas dissolved in it, made an orange dye. Logwood chips steeped in

*Many Scottish immigrants were accustomed to dyeing and cloth work. James Boswell observed in 1773 that the inhabitants of the Isle of Coll 'can all dye. They use heath for yellow; and for red, a moss which grows on stones. They make broadcloth, and tartan and linen, of their own wool and flax, enough for themselves, as also stockings.' (*Journal of a Tour to the Hebrides with Samuel Johnson, LL.D.*)

an iron pot in cider or vinegar with copperas provided a good black, the yarn being boiled in it for half an hour and then rinsed many times in cold spring water before drying in a shady place. Yarn was sometimes clouded various shades of blue by braiding three skeins of yarn before dipping in the indigo vat. To make the yarn light and dark alternately, tight bands of cotton were wound equidistant around the skeins before dipping in the dye tub.

Faded colours could sometimes be brightened by rinsing in the right solution. A pearl-ash solution improved faded purples or lilacs, while a little vinegar in the water restored reds and pinks. There were, of course, numerous dyes for sale at the apothecary's shop, but most settlers avoided purchases of any kind whenever possible. Many simple dyes were prepared in the kitchen, but some farmers who entered upon the work on a larger scale built a small shed for the purpose. The dye-house, a place of evil smells, was on that account usually erected at a distance from the home, and most settlers were glad enough to give up the business to the skilled dyer when he appeared in the community.

(5) WEAVING

The weaving process was carried out on two types of looms —the hand-loom for braid and ribbons, and the large loom for heavier cloth. The girls usually filled the quills with yarn, and these held it in the shuttle. The yarn was stretched upon pins to form the warp, while the woof or weft was carried under and over the warp at right-angles to it by the weaver or a mechanical shuttle. The loom held the threads tight, straight, and parallel, raising alternate threads and lowering those intervening. The bobbin was a small spool held in the weaver's hand and from which the thread unwound; and as he wove he pushed the new weaving tightly against the old with his other hand.

It took practice to weave a smooth even cloth, and most settlers did not attempt it. Even the crudest of looms was complicated, while the power looms in woollen mills were elaborate machines. Many mills were established in Upper Canada in the 1840's and 1850's. Usually two pounds of wool made a yard of cloth, and to the smaller factories a farmer might take his crop of wool and receive back half of the resulting cloth. The larger plants usually purchased their wool and sold their entire production of cloth.

Domestic weaving was more general in Lower Canada than in the upper province, and many thousands of looms were in use there.

In the month of October [says one of the pioneers] great webs of this homespun could be found in the house of almost every settler. . . . There was very little fine goods at that time, and the agriculturist, when he was 'togged out' in his homespun, and his boots had got an extra dose of tallow, considered himself 'nae sheep-shank'.

One of the most primitive examples of ingenuity in making cloth is provided by a Loyalist woman. She lived near a small tannery and was given the hair from the tanner's vat. She cleaned, carded, and fulled it, and combining it with a hemp-like weed growing near her house, she doubled and twisted it into thread and wove it into her only blanket.

(6) FULLING

The fulling process, like the carding of wool, was often left to the factory rather than the home worker, but a fulling bee was called if the farmer intended to full his own cloth—that is, to make it shorter, thicker, and stronger. Long bench-tables were placed around the room for the cloth, and the guests at the frolic sat around them at intervals sufficient to allow each worker full use of his arms. The long web of cloth, which had been well soaked and soaped, was laid on the table, and every boy and girl seized a portion and pulled, pushed, and twisted it with great vigour. As they worked, they kept the great web gradually moving around the circle from right to left. Perhaps someone would start a song, which, like a sailors' chant, added to the jollity and enabled the fulling to be done with rhythmic precision* Work was forgotten as jokes and flirtation whiled away the time agreeably. After a couple of hours the cloth was fulled, and its greater strength and durability would long bear witness to the effectiveness of the workers. The web of cloth

*The thickening bee has come down through the centuries. A traveller in Scotland nearly two hundred years ago observed the *wawking* or thickening of cloth: 'Twelve or fourteen women, divided into two equal numbers, sit down on each side of a long board, ribbed lengthways, placing the cloth on it: first they begin to work it backwards and forwards with their hands, singing at the same time, as at the quern: when they have tired their hands, every female uses her feet for the same purpose and six or seven pair of naked feet are in the most violent agitation, working one against the other: as by this time they grow very earnest in their labours, the fury of the song rises; at length it arrives to such a pitch that without breach of charity you would imagine a troop of female daemoniacs to have been assembled.' (Thomas Pennant, *A Tour in Scotland and Voyage to the Hebrides,* 1772, pp. 285-6.)

was carefully smoothed out flat on the tables, and then laid aside to be dried next day. A hearty supper and a still heartier square dance to the fiddle concluded the bee.

(7) MAKING CLOTHES

The fullcloth made fine suits and dresses, and gay home-spun was long popular and almost universal in the rural districts. Perhaps the flannel had been glossed at a fulling mill, for a glossy flannel, we are told, was 'an ideal for most girls, that they hoped to attain some day in the far distant future'. Some girls were so skilful at weaving that they could weave ten or twelve yards of fine flannel a day. A pioneer of the Talbot Settlement describes the great exertions and extensive travelling he had to undertake to make eighteen yards of flannel:

In the year 1813 Colonel Talbot sent word to the few settlers that he had wool to be let out to be made into cloth on halves. I hired a horse and went and got fifty pounds. Here was forty miles travelled. I then hired a horse and took the wool to Port Dover and had it carded, for which I paid $6.25 and returned home, which made one hundred miles more. My wife spun the rolls, and I had made a loom for weaving but we had no reed for flannel. I then went sixty miles on foot to a reed-maker's, but he had none that was suitable, and would not leave his work on the farm until I agreed to give him the price of two reeds, $6.50, and work a day in his place; this I did and returned home with the reed. My wife wove the cloth, and I took my half to Dover to the fulling-mill. When finished I had eighteen yards, for which I had paid $34.75 and travelled 140 miles on horseback and 260 miles on foot, making four hundred miles, requiring in all about fifteen days' labour.

Wool clothes were preferred to cotton for a variety of reasons. In winter they kept out the cold, and in summer absorbed perspiration; while the busy housewife often found that cotton dresses were readily set afire as she worked about the hot stove. A fine woollen shawl, worn over the head like a hood, was popular for winter. In making clothes, perhaps the buttons alone were purchased, and one set often lasted for several suits. Most country-people made the everyday clothing of their menfolk, the pattern being obtained by ripping to pieces some old garment of good make—perhaps one which had been devoted to Sunday wear in the Old Land. Even if these home-made garments were not sewn with the skill of a tailor, they well served the purpose. Perhaps a man had one suit for church wear cut out by a town tailor—possibly even made up by him;

J. E. Laughlin

SPINNING AND WEAVING

Harold McCrea

THE QUILTING BEE
A great clearing-house for gossip.

and ready-made clothing could often be purchased in the general store.

At first the women made most of their own clothing. In early Loyalist times even a calico dress was a luxury set aside for a girl's wedding day, but as the years passed, more and more clothing could be purchased. Dress-makers and tailors frequently went about the country, making a year's supply of clothing while they boarded in the home—generally using the homespun which had been prepared for their coming.

(8) QUILT-MAKING

The weaving of blankets, quilts, and counterpanes was especially enjoyed by the women. The wool had, of course, been dyed, and the geometric pattern carefully worked out. Some of the favourites were known as 'Lovers' Knot', 'The Rose', 'Five-Star', and 'Chariot Wheels'.

In making patchwork quilts, pieces of cotton or wool cloth were arranged in 12-inch squares, which were sewn together to form the top of the quilt. Many were the familiar patterns—the 'Log Cabin', the 'Flower Basket', the 'Double Irish Chain'— and the skill in stitching and arranging colours is apparent in the quilts still proudly displayed in many homes. During the quilting bee the top of the quilt and its lining were fastened to a frame and quilted together, each woman working on a section; and the edges were finished with a binding.

(9) KNITTING AND EMBROIDERY

Knitting was one of the most general crafts among pioneer women, for there was a great demand for warm wool socks, stockings, mitts, and long gauntlets during the cold winter. Surplus socks could always find a market at the general store. One girl knitted seventy-five pairs of socks and with the proceeds purchased her trousseau, excepting such clothing as she could knit herself.

Ingenious designs were worked into mitts and socks by the best knitters. Some were striped, some spotted, others flowered or plain, ribbed or unribbed. Cradle quilts, chair covers, coverlets, and curtains were other products of the knitter's skill, and it was an early social custom for girls to carry their knitting with them when paying calls on their neighbours. The domestic

SAMPLER WORKED BY CHARLOTTE PAYNE

Many interesting descriptions of samplers may be found in Ethel Willson Trewhella's pamphlet *The Romance of Samplers: Being an Account of Their Origin, and Telling of Several which May Be Found in Newmarket and District.*

arts exhibits in many a rural fair showed the careful workmanship and unremitting perseverance of the women.*

Closely related to work in wool was embroidery in cotton or silk. Many a young girl exemplified her skill in embroidery in making a sampler—in fact, it was almost a religious duty to do so. Coarse canvas cloth was generally used as a background, and on it the alphabet, numerals, the Lord's Prayer, and Bible verses were carefully worked amid pictures, flowers, and designs. Usually many kinds of stitching were employed, and when finished the sampler was framed. As the maker's name and age was always sewn upon it, the sampler is a prized heirloom in many a home.

Prior to Confederation, efforts were made to introduce the silk industry in Upper Canada. Mulberry trees were grown in Hamilton (Mulberry Street commemorates the locality), and a group of women of St. Catharines exhibited native silk at the Provincial Exhibition in Hamilton. But the Canadian climate was not sufficiently suitable for the industry to develop.

(10) STRAW WORK

The husks or straw of corn, oats, wheat, and rye were used by economical and industrious settlers for several purposes. Corn husks were preferred to straw for filling mattresses, but a nail comb—made by driving six or eight nails through a board—first tore them into thin strips. As they fell into the coarse linen bag, they were stirred and arranged by hand to make a soft and even mattress.

Stalks of grain to be used for straw weaving are usually cut before quite ripe, and so are not too brittle. The pioneers found rye straw most pliable, and that of the wild rice was also good. The straw was first braided into strands, and these were sewn together to make straw hats, baskets, etc. A wooden or iron needle united the individual strands of the ropes, though continual threading and tying proved too monotonous for some, who avoided it by using twine or the tough inner bark of the cedar to join the ropes of straw. Mrs. Traill gives the following directions for making door mats out of corn sheathing:

*One of the best early books on the subject had some circulation in Ontario. It is *The Ladies' Hand Book of Fancy and Ornamental Work, Comprising Directions and Patterns for Working in Appliqué, Bead Work, Braiding, Canvas Work, Knitting, Netting, Tatting, Worsted Work, Quilting, Patchwork, &c., &c.* Illustrated with 262 Engravings. Compiled from the Best Authorities, by Miss Florence Hartley. Philadelphia: 1861.

The rough ends of the husk are left projecting about an inch. The braid is made in this fashion: you take nine blades of the sheathing and tie them at the top to keep your work from coming undone: the braid is the simple three ply; but you use three blades together instead of one. To make it thick enough, every time you come to the left side insert there a fresh blade, leaving a little bit of the end to project at the edge. About twenty yards is sufficient for a door mat; it is sewn together with the big needle and twine or bark. Children can be taught to make these things; and they cost nothing but the time, and can be made of an evening, or on wet days when other work cannot be attended to.

Hats or mats of straw might be bleached by exposure to sulphur fumes in a covered box or barrel, but the mats were usually left in their natural colour. Some farmers made beehives of straw ropes, commencing at the top of the cone and working downward. The coils were bound together by twine or strips of elm or cedar bark. The bees were believed to aid in making their home still warmer by sealing up any holes. Among the various types of straw baskets were conical receptacles used by the 'Pennsylvania Dutch' immigrants for raising bread. In some districts the inner bark of the basswood and oak were used in weaving baskets and mats.

(11) RUGS AND CARPETS

Closely related to the weaving of straw mats is the making of carpets from yarn or rags. The coarser yarns, dyed in gay colours, were suitable for carpeting, but most people merely assembled the dyed yarn and sent it to weavers to be made into carpets. Rag carpets were, however, quite easy to make. Usually they were vivid in colour, but the pattern varied from none at all to intricate and beautiful designs, which formed a startling contrast to the otherwise drab log house.

The first work in rug-making was to assemble, cut up, and wind into balls all sorts of old clothes—linen, cotton, and woollen. A careful observer saw a pioneer woman with a basketful of such materials in front of her, and describes the work:

She then took a piece, and with the scissors began cutting it into long narrow strips about a quarter of an inch wide, not wider; and indeed the narrower the strip the better. She did not cut quite through when she came to the end, but left just as much as would serve to hold it together with the next strip, turning the piece in her hand and making another cut; and so she went on cutting and tearing, till that piece was disposed; she then proceeded to a second, having first wound up the long strip: if a break occurred she joined it with a needle and thread by tacking it with a stitch or two. Sometimes she got a bit that would tear easily, and then she went on very quickly with her work. Instead of selecting her rags all of one shade for the ball, she would join all kinds of colours and materials. 'The more lively the contrast, the better the carpet would look,' she said.

Sometimes, however, each colour was wound in a separate ball, and a striped or chain effect resulted when the various colours were woven into the rug in turn; while by the other method a hit-and-miss mottled pattern was the result.

Very large looms were needed for weaving rag carpets, so almost all settlers sent a certain number of pounds of these large balls to the weaver, together with cotton warp which had been doubled and twisted on the spinning-wheel. The weaver's charge was about 6d. a yard, and a pound and a half of rags, with the necessary warp, made a yard of carpet. Sometimes the warp was dyed by the owner, sometimes by the weaver.

Children were often taught to cut and join rags and wind them. For the best result, the rags were cut right through instead of turning the corners, for the carpet was much more regular, though more sewing was necessary. As even ordinary carpets cost 4s. a yard, a considerable saving was effected by the use of waste materials.

In modern times the inhabitants of Quebec are noted for their home-made hooked rugs. They are not woven on a loom, however, but worked with a hook on canvas stretched over a wooden frame somewhat similar to a quilting-frame.

(12) THE CULTIVATION OF HEMP

Flax and hemp, the fibres of which are used to make linen, sailcloth, ropes, etc., were in demand in the Canadian pioneer period. Premiums were offered by the Government for their cultivation in the British Provinces in America so that the British navy would not lack a continuous supply of sails and ropes. But, in spite of this encouragement, neither hemp nor flax was ever generally cultivated in Canada.

Hemp adapts itself to the hot climate of the tropics as well as to the rugged north temperate zone. It grows to a height of from three or four to twenty feet or more, the female flower on higher and stronger plants than the male. The stems are either hollow or filled with a soft pith, with a woody fibre and then a bark surrounding it. It is this outer fibre that the growers remove.

The more rapid the growth of hemp the longer the fibres, and the coarser fibres result rather from thin sowing than thick. Soon after flowering, the plant is pulled out by hand, and its treatment thereafter is similar to that of flax. While some of the finest fibres are similar to flax, they are in general used for coarser purposes, such as the manufacture of sailcloth and ropes, and for the caulking of ships. Oil of hempseed is used in some countries as fuel for lamps, as well as in the manufacture of paint, varnish, and soft soap.

Many early settlers attempted hemp culture, but with varying success. Philemon Wright, 'the White Chief of the Ottawa', was among the most successful cultivators, growing eighty-five per cent of the total in Lower Canada in 1802 and being awarded a silver medal by the Society of Arts. His crop grew to a height of fourteen feet, and he considered that Canada was a fine country for hemp. But the expense of preparing it for market led him—and most other growers—to give up its cultivation on a large scale, though he continued to grow small quantities for his own use.

My hemp-peelers [he wrote] charged me one dollar per day, or one bushel of wheat, labourers being very scarce in the township of Hull. I sowed nearly 100 bushels of hemp-seed, which I sold in Montreal at a fair price. I was obliged to send the hemp to Halifax to find a sale for it. . . . I also built a hemp-mill, which cost me £300, which mill was by accident burnt.

Along the seacoast there were ropemakers who wove the strands of hemp fibre into ropes by the use of wheels, walking along as they twisted the fibres.

(13) THE CULTIVATION OF FLAX

Flax cultivation was more extensive, particularly in Lower Canada, where 1,313,648 lbs. was produced in 1827. The seed was sown in damp locations during the last week of April, and by the middle of July it had blossomed and was about to go to seed. It was then ready to pull, stem and root, which was best done by hand to avoid injury to the fibres. In later years a special machine for the purpose was introduced, while many farmers found that a reaper, if it cut close to the ground, did the work satisfactorily for all but the finest linen.

The pulled flax was bound in small sheaves and spread out on the ground to be cured in the sun. The *bobs,* or seeds, were taken out, to be used for the following year's seeding or sold

Courtesy of the Artist   Jerrine Kinton
FLAX, WATERLOO COUNTY

to an extractor of flaxseed oil. At the middle of the nineteenth century there were numerous factories where oil was extracted and oil-cake made. The process of removing the seeds was called ripping, and was done by a comb of heavy wood or iron, with which the flax was beaten. The removal of the leaves was called retting.

The most difficult processes in the preparation of flax were yet to come, and many an unskilled farmer failed to perform them properly. The straw had to be split into fibres, and the useless centre taken out, but a certain amount of weathering and soaking was first needed to rot the pith. This *steep* lasted from ten to twenty days, and a process of fermentation dissolved the cement which holds the filaments and fibres together, enabling the removal of the *shave* or woody centre.

The next step was the use of the swingle-board or scutching machine. The flax was pounded between sets of long wooden knives or flax brakes on a heavy block of wood: the worker held a bunch of flax in his left hand and brought the brake (or *break*) down with force with his right. It was hard and dusty work.

The hackling or hetcheling followed. In this process the good silky fibre was separated from the coarse tow or *shives* by combing the slightly moistened fibre. A long wooden *sword,*

Conservation Authorities, Ontario

HAND-MADE MACHINE FOR BRAKING FLAX STALKS

sharp on both sides, was sometimes used, or the large prickly head of the teasel or *fuller's thistle*. The hackle teeth straightened the fibres, and fineness of fibre resulted from much hackling. If these processes were successfully carried out the farmer had a soft, silky fibre ready to spin.

The difficulties of preparing hemp and flax for the spinners and weavers indicated the need of mills for the purpose. But an emigrant guidebook of 1843 states that there was not then a single flax mill in Upper Canada. It was therefore suggested that groups of neighbours should combine to erect a mill in which their flax and hemp could be treated. This was not often done, however, and many settlers discontinued cultivation rather than attempt to surmount the difficulties.

(14) MAKING LINEN

A smaller wheel was used for spinning linen than for spinning yarn. The distaff was a forked piece of wood at the front of the wheel, and on it was placed the whitened flax fibre. The spinner pulled it off and wrapped it around the spindle, turning the wheel by a foot treadle: the fibre was lengthened out and twisted into thread, which was then wound on a bobbin. When the bobbin was full, the linen thread was wound or tied into skeins or *hanks,* usually by one of the daughters using a reel.

To make the skeins white a bleaching process followed. Ashes, slaked lime, or sour milk formed the usual solution for the purpose, but the best results were obtained only by twenty or thirty insertions, between which the skeins were dried in the sun.

Often a year and a half elapsed between the planting of the flaxseed and the finished linen, for the thread had still to be woven. The loom used for woollen cloth was similarly employed for weaving linen, the earliest Loyalist settlers using a mixture

Canada Farmer

FLAX MILL AT STREETSVILLE, 1866

to produce *linsey-woolsey.* The *tow,* or coarse fibres, were commonly made into ropes, grain bags, or rough working trousers, but the best made fine linen sheets. At first the cloth was hard and stiff, but continual washing and pounding worked it into a softer texture. This was heavy work, for all the grease had to be removed, and the boys and men generally did most of it. In Lower Canada a frolic was sometimes held from house to house to do the fulling, boys and girls dancing up and down upon the cloth in big tubs until it was soft.

Mrs. Sarah Slaght, an early settler in the Long Point district, tells with pardonable pride how she made her supply of linen during the first year of her marriage:

In the barn there was a quantity of flax. . . . Job broke it for me and then I took off the shives, hetcheled it, takin' out the tow, which was carded and spun on the big wheel like wool, and which furnished the fillin' in weavin' the coarser cloth used for towelling, tickin', bagging, etc. The flax was spun on the little wheel. We were married in October, and during the winter I made up forty-three yards of cloth out of that flax, and this gave us a supply of table cloths, towels, sheets, tickin', bags, etc.

Besides the difficult process,* the extension of the flax industry was retarded by the lack of a market for the straw. But cultivation increased in the eighteen-sixties, especially in Lower Canada, for there was a scarcity of cotton during the American Civil War. In 1861 the linen produced in Upper Canada was 37,055 yards, but in Lower Canada** the production totalled over one million yards. Improvements in the process of manufacture included soaking the flax in water of a temperature of 80° to 90° instead of the long weathering, but this was not commonly done in Canada.

*The Ottawa River, John Mactaggart noted, would bleach coarse linens like Irish Dowlas very effectively. Whether from vegetable or mineral matter in the water or the strong summer sun, he found that 'without any chemical process, without any artificial alkalis, here we see the dingy web becoming like snow in the course of a couple of weeks'.

**The names Upper and Lower Canada were commonly used long after the provinces became officially known as Canada West and Canada East by the terms of the Act of Union, 1841.

(1) MAPLE SUGAR MAKING

LIKE the construction and use of the birch canoe, the manufacture of maple sugar is one of the contributions of the American Indian to our civilization. In the early seventeenth century French missionaries observed the Indians making sugar each spring in a primitive manner. In the Jesuit *Relations* there are references to 'Maple Water' running from the trees 'toward the end of Winter'; and to the tree Micktan, 'which they split in the Spring to get from it a juice, sweet as honey or as sugar'. In times of famine the Indians were observed eating the bark of this tree—the sugar maple.

The Indian process of manufacture was crude. The tree was gashed with a tomahawk, and a wooden chip carried the sap drop by drop into a birch-bark receptacle; or, in the case of the Mohawks, into a basswood sap trough hollowed by a stone adze with the aid of fire.

In boiling the sap, earthenware pots were used by those who possessed them; but some tribes had no better method than to drop red-hot stones into the sap trough, continuing this crude procedure until the sap was eventually boiled down. This laborious and primitive process produced only a small quantity of coarse and dirty sugar, but even that was highly prized and formed a considerable part of the food supply of some tribes. With the coming of fur trader and settler, the Indians obtained iron kettles and carried on the process in a more civilized manner. In the sugar season they usually encamped near a maple grove, where the women of the tribe entered into the work. The sugar was usually packed in neat birch-bark baskets called *mokuks,* and at early markets these boxes were a characteristic sight, the usual price of the sugar being from six to ten cents a pound.

The settlers carried on the industry in much the same fashion. Sometimes sugar was also made from the sap of the black walnut. Mrs. J. G. Simcoe, wife of the first lieutenant-governor of Upper Canada, was given a sample, and found that it 'looks darker than that from the maple, but I think it is sweeter'. While sugar-making was heavy work it was commonly a time of enjoyment, for it was in process during the season which heralded the end of winter. To ensure cleanliness and convenience the sugar-bush was carefully underbrushed and rotten logs and

MAKING MAPLE SUGAR

other obstructions removed. If cattle were in the neighbourhood a fence was erected to keep them away from the sap-troughs.

Careless settlers merely gashed the tree with an axe to set the sap flowing, but the proper method to avoid permanent damage to the sugar-bush was to make a hole with an inch auger. Sometimes the gash of an axe produced the better flow, but unless it was intended to cut the tree down shortly it was much better to use an auger or a tapping iron. A round spile, hollow in the centre, was driven about an inch into the hole, or a gouge was made below the hole to hold the gutter. It was found that the tap was best placed on the south side of the tree in the early weeks of the season, but on the north side if it required renewing towards the end. The sap ran best on warm sunny days following frosty nights.

Nowadays pails are commonly used to hold the sap, but sap-troughs of pine, ash, cherry, or butternut, holding three or four gallons each, were more generally used by early settlers. The troughs were placed directly under the drop from the spiles, and at regular intervals the sap was collected.

At the centre of the grove a place was selected for the boiling, and roads leading to it radiated in all directions. Sometimes the men carried the sap in pails by the use of a wooden collar or yoke about their shoulders, from which cords were suspended; but large-scale producers frequently used an ox-sleigh carrying one or more barrels of sap to the boiling-place. A wooden tundish or funnel with the capacity of a sap-trough provided a convenient aid in pouring the sap into the barrel. A large store-trough at the boiling-place held the sap until all was in readiness for boiling down, and extra barrels were provided if the flow of sap was exceptionally good.

Meanwhile the children have aided in collecting and preparing a large supply of firewood, for when boiling commences, the fires must burn steadily for many hours—often right through the night. As evaporation proceeds, the sap becomes a thin syrup and is almost ready for the 'sugaring-off', a festive time when sugar-eating bees and a dance to the fiddle provided enjoyment for the boys and girls. But first the molasses was carefully strained through wool into a copper boiler, and a clarifier used to remove earthy particles and other impurities. Milk or a piece of fat pork was sometimes used for the purpose, but eggs were more satisfactory. Half a dozen eggs, with the shells, were beaten up with a quart of the molasses, the mixture was stirred into the kettle while both were cool, and the pot was then brought to a boil over a slow fire. As it began to simmer, the impurities rose to the surface with the beaten eggs, and just as the boiling-point was reached the kettle was swung off the fire and the scum carefully skimmed off, leaving the molasses bright and clear.

Again the liquid was placed over the fire, and this time it had to be even more carefully tended, for it would boil over in an instant—a disaster which, however, might be halted by pouring a little new sap into the pot or throwing a shovel of snow on the fire. Several methods were employed to detect the exact time to remove the boiling syrup from the fire. If it

MR. AND MRS.
JAMES O'HARA
Golden Wedding
Anniversary,
February 3,
1898.

Conservation Authorities, Ontario

SUGARING-OFF, MOIRA RIVER

hardened at once when dropped on snow it was ready. Samuel Strickland used a stick on which he had cut a narrow slit an inch long and one-eighth of an inch wide. When this was dipped into the molasses at the right time a thin film would fill the hole; then, if a long bubble could be blown from it, the molasses was sufficiently boiled. Pans or moulds received it, and there it hardened into cakes.

Besides maple syrup there was maple butter, which was in consistency about half way between syrup and sugar. Mrs. Traill sometimes made 'maple sugar sweeties' during the sugaring-off. These were small candies formed by pouring thick syrup into a dish, stirring in a little flour and butter, and flavouring with lemon, peppermint, or ginger. When cold and hard the candy was cut into small squares.

There were many variations in the process of sugar-making. On some farms several kettles were used in the boiling and the sap poured from one to another as it got thicker, fresh sap being added to the first boiler from time to time. We are told that sometimes the mother of the family supervised the boiling, rocking her youngest baby in a sap-trough cradle meantime.

The early sap was found to be much the best. April sap soured quickly and was difficult to grain into sugar. A little saleratus or a handful of quick lime would remove some of the acidity, but such sap was best used for vinegar. While maple syrup* was not extensively made in the pioneer period, most settlers who had a maple grove produced large quantities of sugar, for West Indian sugar was expensive. A sugar-bush of 500 trees produced from 800 to 1,200 lbs. of sugar in a good season, and many farmers had from 1,000 to 3,000 lbs. as a result of their work. Nowadays the greater part of Canada's maple sugar is produced in Quebec and exported to the United States to be used in curing tobacco.

(2) LIME-BURNING

Commercial lime is obtained by burning limestone in a kiln. The resulting quicklime, or burned lime, may be turned into slaked lime by the addition of water, which combines with the quicklime to form a light, white, dry powder; and this, mixed with sand or used alone, becomes mortar by gradually hardening as it absorbs carbonic acid from the air.

*William Adair recalled making 70 gallons of maple syrup and taking it to London to sell. He sold one gallon for 50 cents and took the rest back home.

SOAP MAKING

OLD LIME-KILNS NEAR KELSO, NASSAGAWAYA TOWNSHIP

Lime-burning was a process of importance to the pioneer settler when he had prospered sufficiently to improve his log shanty. Many regions, however, contained no limestone, while settlers near towns often preferred to buy their supply.

The easiest method of burning limestone did not produce the whitest lime but was more commonly used. Sometimes lime was burned while logging was in progress, for a very large log-heap was needed. Samuel Strickland calculated that he would need about one hundred bushels for plastering his walls and building chimneys, so he constructed an immense pile of the largest logs available. The timber from half an acre was needed to build the heap, and on top a framework of logs was constructed to hold the limestone.

Twenty cartloads of stone were then drawn to the log-heap and placed within the frame after being broken into small pieces with a sledge-hammer. The fire was then lit, and the pile burned during the night; but the intense heat of the coals prevented the lime being removed until a week had passed. It was then collected and covered, but it was impossible to eliminate entirely small pieces of unburnt stone, which detracted from the value of the lime as plaster.

As settlement progressed a few men in each district where limestone was available specialized in lime-burning. Many villages had two or three large kilns which supplied lime to the surrounding territory.

Some settlers used clay as a substitute for lime. To prepare it for use the clay was mixed with boiling water and then worked in front of a fire. The resulting material made a satisfactory cement.

(3) CHARCOAL-BURNING

Charcoal-burning has not been as extensively practised in Canada as in the Old World, but some pioneer farmers supplied themselves with charcoal. The process consists of burning wood in a limited supply of air, which has the effect of retaining most of the carbon, instead of burning the wood to white ashes in the open air.

The charcoal-heap is usually of conical form, consisting of broken pieces of wood in rows. Holes are left at the bottom for the entry of air, and a hollow space in the centre to serve as a flue for the escape of gas. Over the heap a covering of ashes or turf is placed, and the pile is ignited by throwing burning pieces of wood into the central opening, where a wooden grate prevented their falling at once to the bottom.

In making a 'pit' of charcoal [wrote a pioneer producer of it, William Adair, who lived on the county line between Middlesex and Elgin and sold his charcoal for a York shilling—12½ cents—a bushel] about ten to twelve cords of wood were used. A pole was set in the ground. Cordwood was piled on end in two tiers, with shorter pieces to round off the top. Straw was piled all around the wood, with air holes around the bottom. The fire was started at the top and then the top was covered with earth. It took from three to five days to burn a pit. . . . The men stayed at the job day and night. The pit gradually decreased in height. At last some coals, on the lowest side, were raked out, and soon there was a circle of black coal around the pit.

The burning of a charcoal-heap proceeds gradually from top to bottom and to the outside, and fresh wood is thrown in continually as the central portion burns away. If the smoke is thick and white the combustion is satisfactory, but if it is thin or a blue flame appears, the wood is burning too fast and the air-holes must be partially stopped up and more sod placed on the heap. When the burning is complete the whole is covered over and left to cool for two or three days; whereupon the heap is broken up and water thrown on any pieces still hot. Usually the volume of charcoal is about three-fifths of the wood used, or by weight about one-quarter. The uses of charcoal are numerous, but for the pioneer settler it was merely a first-class fuel that he obtained. John Thomson, settled near Lake Simcoe, noted in his diary on several occasions that his men were engaged in the work. 'Five hands commenced cutting basswood logs and splitting them to make a charcoal heap,' he wrote on October 14, 1834. On another occasion he noted that his men were tending the charcoal-heaps both night and day.*

(4) MAKING SOAP

During the clearing of forests for pioneer settlements most of the timber was burned to get rid of it as speedily as possible. In many instances, however, an effort was made to save the best ashes, for they were extensively used in the manufacture of soap and fertilizers at that time. If it was not possible to use

*Two Canadians of central European origin developed charcoal-burning near Apsley, Peterborough County, in the nineteen-forties. They found a ready market for their product in the city of Peterborough.

them immediately cribs of logs were built to store them, and
large strips of bark served as a covering. Hardwood ashes were
much the best for soap—in fact, many softwood trees were
entirely useless.

Grease was the other main ingredient which entered into
the manufacture of soap. Most pioneer homes had consequently a soap-box where all sorts of fat rinds, dripping, grease, etc.,
were stored in readiness for soap-making. The entrails of animals were used, and even bones were boiled down in strong
lye, the lime content improving the soap. Enough soap was
commonly made to last a year, and it was thought best to
choose a time when the moon was new, for the soap tended to
dry out too much as the moon waned.

At first ash-leaches were hollowed basswood logs set on end
on a drainage platform of split slabs, but barrels or vats were
used as they became available. The leach was slightly tilted on
a grooved board or trough and raised a couple of feet above
the ground, so that a crock or tub could be placed under the
groove to catch the lye.

At the bottom of the leach a layer of split lathing or twigs
was placed, and then a layer of straw. Sometimes a layer of
lime was placed next, or it could be dissolved in boiling water
and mixed throughout the barrel or even poured in at the top,
the total amount being about two quarts to a barrel of ashes;
its purpose was to neutralize certain salts which were injurious
to the soap. The ashes were then put in and pounded down,
and at the top a round hole was made to receive the water. Hot
water was first put in, and then as it filtered through, cold soft
water was added from time to time. When the ashes were
packed tight, it was sometimes two or three days before the lye
began to seep through the holes into the groove beneath. As
the water soaked through it dissolved the alkali in the ashes,
and the lye ran out into the receptacle beneath. When the lye
was seen to become weak and light in colour it was apparent
that all the strength had been extracted from the ashes. The
lye was sometimes tested for strength, being about right when
it would buoy up an egg or potato half above the surface. A
personal recollection adds a good deal to the record:

One of my boyhood memories [writes the Reverend L. F. Kipp, long
a prominent journalist in Montreal and Toronto] was the annual
making of soft soap, and the appearance two or three times per year

of the ashman. Our wood ashes were put into a little shed near the orchard fence.

As far as soft soap was concerned, it was quite an operation. On a platform grooved to run the lye down to a great iron kettle was a 4-foot log hollowed out to the platform. This was filled with ashes, and pails of water poured into the top. In a few days a deep-red liquid came out. When mother had sufficient, a fire was set under the big iron kettle into which went all the bones and fat saved in the winter. When the lye had eaten as many bones and fat as it could you had a soap like liver. Believe me, it would take off any dirt misplaced—and the skin on your hands as well!*

While 'cold soap' could be made without boiling, in general a soap kettle was needed to produce a satisfactory result. Crotched sticks a few feet apart in the ground supported a pole which suspended the kettle by a chain, and underneath a steady fire boiled the lye and grease. Some used twelve pounds of grease in making a barrel of soft soap, others three pounds to a pail of strong lye. When there was a thick scum of grease on top the lye had absorbed its quota, and more lye was poured in; while if the soap did not thicken and there was no scum at the top more grease was added. For soft soap the kettle was removed when the contents formed a clear mass, and a crock received the slimy material. It stood by the family wash-basin and you helped yourself to a handful of soft soap when you washed.

Sometimes the soap was too strong and needed water to make it thicken. By testing a few spoonfuls in a saucer the proportion of water to add was found; in some instances, where the lye was very strong, as much as double the quantity of water could be added. A pint of turpentine or some resin was sometimes poured in during the boiling to improve the quality of the soap. A barrel of ashes and twelve pounds of grease would produce some forty pounds of soap, and the process was usually completed in one day. Many an elderly man, unwilling to change habits of long standing, continued to make soft soap for his own use, while his children and grandchildren preferred 'new-fangled' things.

'Cold soap' could also be made in a barrel through the action of the sun. The grease was boiled down in weak lye and strained into the barrel, being mixed with strong lye and stirred from time to time. If the mixture did not thicken in a week or ten days more lye or grease was added. Some settlers

---

*Letter to the author, January 30, 1962. Mr. Kipp is 84 years of age, and his vivid memories go back to the mid-eighteen-eighties.

made a low-grade soap by merely mixing the fat and lye, while others carried out part of the process by boiling the grease and lye and then letting the sun do the rest.

Hard soap was made from good soft soap, or by continuing the soft-soap process a little further. Several handfuls of salt were stirred into the boiling soap, and as it boiled, it was seen that the soap was coming to the top and the reddish-brown lye sinking to the bottom. Set aside overnight the soap formed a thick cake, which was removed and boiled with turpentine or resin and more salt. When it thickened it was ready to pour into wooden moulds or shallow pans, and as it got hard it was cut into bars and placed behind the stove, where the heat dried it out.

Numerous washing mixtures were made by boiling soap with quicklime and sal-soda (washing soda), and adding resin, turpentine, and salt. Potash soap was another variety of soft soap, made in much the same fashion. Washing-boards, with wooden or zinc rollers, were soon found in the homes of the well-to-do, for, as Mrs. Traill put it, 'In Canada, no servant will wash without a washing-board.'

When lye was extracted on a large scale the soap manufactory was usually located on a running stream, from which water was pumped and distributed to the leaches by wooden gutters or pipes.

(5) THE POTASH INDUSTRY

It was early found that North American ashes contained a larger percentage of pure potash than those of the Baltic regions, and consequently there was long a steady demand for American ashes. As the potash industry was somewhat intricate most settlers did not proceed beyond the extracting of the *ley* or lye, or at most to the point where they obtained black salts. Those who undertook the complete process found that considerable equipment, knowledge, and skill were essential.

There was a great difference in the amount and quality of ashes, and of the resulting potash, which could be obtained from various plants and trees. Ferns furnished most ashes, shrubs gave more than trees, and leaves and small branches more than trunks. The hardest woods, and particularly beech, provided most alkali, while pine and other soft woods were generally unsuitable. There were, too, variations in the process for different types of ash. To obtain satisfactory results earth and other impurities had to be kept out of the ashes, and there must be no unburnt particles of wood. For good ashes farmers could obtain from 6d. to 9d. a bushel at a potashery.

Large iron kettles, costing from $80 to $100 each, were made at Three Rivers especially for the potash industry, for ordinary vessels would not stand the intense heat needed to boil down the lye and melt the salts. Before the lye was boiled down it was sometimes passed through fresh ashes, so that it

Courtesy of Mrs. George Downey

OLD J. B. MACDONALD FOUNDRY, TIVERTON

would have a greater alkali content. When of the right strength it was a dark brown colour, but if it was not clean there was great difficulty in achieving satisfactory results in the rest of the process.

The black salts obtained by boiling down the lye was at first of the consistency of porridge, but as it cooled it quickly hardened like clay mortar. The next step was to melt the salts to eliminate impurities. A farmer with experience in the industry gives a good description of the melting process:

A rough furnace of loose stones was built around the kettle, which was raised a foot or more from the ground so that the fire could be put right under the pot. At the back of this furnace was a chimney of clay and loose stones that carried away the smoke and made the necessary draught. The fire required to melt black salts had to be intensely hot. Indeed, it was so hot that it would nearly melt the potash kettles, big and strong as they were. One had to be very careful when the sticks were poked into the fire, or a hole might be made in the bottom of the pot. . . . Most of this work was done at night, as it was easier to see the colour of the melted stuff, which looked quite a bit like molten iron. When the copper colour was seen—a dark red—the fire was allowed to die down, and the black salts became potash. Sometimes potash was grey when it had cooled, but the best samples were pea-green in colour.

When a blue blaze appeared over the molten mass it was usually poured into another pot called a cooler, which had been heated to receive it. There it became a solid mass and was placed, round side down, in a large oak barrel; another mass, fitting flat side down at the top, made up the characteristic barrel of potash. It was hot, heavy, continuous and dangerous work,* yet pioneer women often took part in it. It is recorded that one woman made eight barrels of potash, at the same time looking after seven young sons and doing most of the farm work.

When potash was made on a large scale it was most convenient to erect the equipment on a hillside along a running stream. Above were the wedge-shaped leaches of double-thick planks, from which the raw lye ran into a reservoir trough or tank. Still lower down was the large potash kettle, perhaps five feet in diameter, into which the lye was dipped by ladles or ran directly through a faucet. Large stones were built around the kettle to hold it in place. At first the boiling took place in daytime, but as the lye became thicker it had not only to be watched day and night but also to be stirred continually with a large spoon shaped like a frying-pan and having a ten-foot handle, of which at least the first four were of iron. If only one kettle was used for boiling, it might take a week to make a barrel of potash, but potasheries usually had several in process at the same time, as well as a series of melting pots and coolers, and ropes and pulleys to handle the solid potash and pack it into barrels for shipment.

Potash was usually shipped eastward to Montreal in large oak-stave barrels holding 560 lbs., which were valued at from $80 to $120 each. The early settler frequently found that cash could be more readily obtained for potash or black salts than for wheat; and in many settlements representatives of potasheries would call and purchase ashes which the farmer had

---

*'I took a job,' wrote Samuel Birdsall in 1862 with reference to 1810, 'of boiling and making potash for the late Captain Humphrey, and worked 22 days and nights, only indulging in bed 4 nights, keeping them [i.e., the fires] going day and night.'

Courtesy of the Artist C. W. Jefferys

MAKING SOAP AND POTASH
Even women engaged in the heavy work.

saved up. A bushel of hardwood ashes made about five pounds of potash, while a 560-lb. barrel was 'the concentrated product of an acre of standing timber, the unbroken forest'. In the 1820's and 1830's the ash trade was one of the most important in Canada. In 1824 about £350,000 was invested in it in Lower Canada—an amount equal to almost half the value of all imports. Canadian ashes were carefully inspected before export and consequently stood high in foreign markets.

Pearl ash was a refined potash, calcined and fused in an oven or a specially constructed furnace. It brought a higher price, was pure white, and was used in making saleratus or baking soda. Sometimes settlers purchased it for baking, for 'a piece the size of your two fists would last the housekeeper for a couple of years for this purpose—it was so strong'. Others burned corncobs to fine white ashes, and from the lye produced an excellent alkali for raising bread.

Potash was used in the manufacture of glass, where it served to clear the sand. It was also used in certain chemical processes, such as making colours fast in printed cotton materials. Until the end of the century a few potasheries remained, but better and cheaper chemical methods of producing potash put an end to this typically pioneer industry.

(6) OTHER FOREST PRODUCTS

Some farmers added to their income by extracting from trees such products as balsam, turpentine, pitch, spruce gum, oil of hemlock, sassafras, and sumach dye. In the Eastern Townships of Lower Canada particularly, there was a considerable trade in hemlock bark, which was used in tanning leather. Just before Confederation the manufacture of wood-pulp and paper from softwood trees became prominent, the aspen or poplar being of particular value for the purpose; this industry, however, was not one for the individual farmer, but required extensive mills.

(1) GRINDING GRAIN INTO FLOUR

IN 1783 the first Loyalist settlers near Kingston took part in a raising bee to construct the first grist mill in the region that, eight years later, was named Upper Canada. This was a government mill, and a similar one was erected the same year near the mouth of Four-Mile Creek in the Niagara district. By 1787 a third mill was in operation on the Napanee River, and before the close of the century a number of privately-owned grist mills had been erected in those parts of the province where extensive settlement was in progress.

But many settlers were in regions remote from any mill, and various substitutes were used to grind grain into flour. Some of the first Loyalists were issued small handmills, in appearance and operation much like a coffee or pepper mill, but difficulty of operation made them unpopular. Later settlers sometimes crushed grain by using stones, after the fashion of the Indians. Another method consisted of boiling grain in lye, and then drying it before the fire until it burst; after which it was boiled in water and eaten as *mush*. This process sometimes took a whole day. The first settlers in Smith Township, Peterborough County, in 1818, had even to resort to chewing corn until it was soft enough for their children to eat. Long trips to the nearest flour mill, sometimes by boat, often on foot through the woods, were a common occurrence in every part of Upper Canada.

The best and most commonly used substitute for stone grinders was the hominy-block or plumping-mill. A hardwood tree-trunk of suitable size was selected and hollowed out by fire, aided by a knife or axe and perhaps a red-hot cannonball, the outside of the stump being kept wet during the process. In the mortar so formed a few quarts of grain was pounded by a hardwood pestle or pounder. Some mortars held a bushel of grain at a time, and a sweep-pole was attached to a branch of a tree so that its resistance, as the pounder was directed up and down, aided greatly in lifting the weighted end. The sweep was most effective if the tree was alive and all the other branches cut off. Two men often worked together at the plumping-mill, but it was not easy to produce good wheat flour in it, though it was satisfactory for cornmeal. Settlers sometimes boiled corn in lye from wood ashes, so that it would burst before being washed, dried, and pounded in the hominy-block.

The coarse brown bran-flour was sometimes put through a horsehair sieve or a piece of cloth, a suitable sifter being passed from family to family. Others winnowed the meal in the wind to remove the husks, and if wind was lacking a current of air was created by waving a blanket. But at best the meal was a poor substitute for white flour as we know it to-day, though undoubtedly producing healthful bread of a coarse type.

Many later settlers in remote regions made long journeys to the nearest mill to get their grain ground, or relied on a coffee mill or some makeshift. Some of the earliest farmers in Eldon Township ground their first grain between two grindstones turned by hand: 'The wheat was poured by hand through a hole in the upper stone. Between dark and bedtime enough would be ground to provide for the next day's needs.' But the 'precious stone flour' from the mill was highly valued in comparison with the *samp* or coarse bran-flour provided by hominy-block or hand-mill.

In 1819 two inhabitants of the Talbot Settlement invented a hand-mill which they called a *bragh*, and within a few months most of the settlers of the district had one. It consisted of granite stones fitted into a framework, the smaller stone on top, and a large bolt passing through the centre of both to fasten them together. A large eye at the top of the bolt made it possible to insert a handspike and carry the mill from place to place. One of its inventors, Peter McKellar, later erected a water-driven mill on Sixteen-Mile Creek, and ingeniously constructed most of its equipment as well.

Grist mills were often the nucleus of thriving villages and towns as remote parts of Upper Canada were opened for settlement, for tavern, general store, and blacksmith shop soon

THE OLD MILL IN WINTER

George H. Durrie

followed, and as the population increased school and church were provided. The grist mill erected by the government on the Apanee River in 1785 became the centre of the settlement of Napanee; Scott's Mills became Peterborough, Purdy's Mills grew into Lindsay, Darlington Mills, Bowmanville, and Shade's Mills, Galt; around Street's mill grew the village of Streetsville, Priest's mill in Glengarry County became Alex-

Conservation Authorities, Ontario   Capt. Thomas Burrowes

NAPANEE ABOUT 1830

andria, and Ward's mills developed into the settlement of Wardsville, which was later named Smith's Falls.

In flat regions, where water power was not available, other means were employed to grind grain. In 1833 Patrick Shirreff visited the French settlements along the Detroit River and observed windmills in general use. He also saw several propelled by oxen walking on an inclined plane, but considered them 'very poor machines'. Others were driven by oxen and horses attached to a large wheel, which moved horizontally a few inches from the ground; and he learned that a steam mill was about to be erected in Sandwich.

The palmy days of the small flour mill were the eighteen-thirties and -forties. In 1836 there were about six hundred in Upper Canada, and in 1854 only a few more. But steam mills were gradually increasing in number, and the ruins of many a small water mill tell a story of inability to keep pace with improvements, such as the turbine wheel, which gradually concentrated flour-milling into fewer and fewer hands. Similarly many a farmer's wife, instead of baking bread, now buys it daily from the baker, whose bread is probably made from flour ground—not from the farmer's own grain—but from the best hard wheat of the Canadian West.

(2) THE PREPARATION OF YEAST

Various types of *barm* or *rising* were in use among early settlers in Upper Canada. Those who lived on farms often grew hops. The right time to gather the cones is when the plant is full grown, with the bright yellow dust completely formed at the base of the blossoms and slightly glutinous to the touch. If left until over-ripe, much of the flavour leaves the hops. Mrs. Traill's hop-rising recipe was as follows:

Boil down two large handfuls of hops in three quarts of water till the hops begin to sink to the bottom of the vessel, which they do after an hour's fast boiling. Put about a quart of flour in an earthen pan or any convenient vessel not too shallow, and strain the liquor, boiling off the fire, into the flour, stirring the batter quickly as you do so. The flour will thicken up like paste: stir it as smoothly as you can, then let it stand till blood warm; mix in a teacupful of the old stock of barm, and let the vessel stand covered up near the fire till it begins to show that fermentation has taken place. In summer you need only cover the jar or pan; it will rise in a few hours; but new barm is not so good as after it has worked for some days. A large earthen pitcher tied down from the air, or a stone jar with a cover, is best for keeping the rising in. The vessel should be well cleaned before refilling.

Another process for hop-rising consisted of boiling a pint of flour and the liquid from hops until it thickened, when a teaspoonful of salt was added and the mixture poured into a jar. A cupful of rising, and sometimes a little brown sugar, was added as it cooled. This type of barm would keep longer without souring. It was best to store all yeast in a cool cellar; but when yeast went sour a little baking soda at the time of use would restore it.

Potatoes, boiled with a little salt and stirred until of the consistency of gruel, were sometimes mixed with the flour. A spoonful of sugar, and two large spoonfuls of rising to start fermentation, were mixed into the liquid before it was strained into a bottle.

Sugar yeast was made by a similar method, but two pounds of soft sugar was used to one pound of flour. This barm would ferment of its own accord and was therefore of special use where no old yeast was available to start the process.

Some settlers made their yeast into cakes. The hop liquor was mixed with rye-meal, and yeast added when cool. Corn-meal was added to make a stiff dough, which was kneaded, rolled to an inch thickness, and cut into cakes, which were left to dry in the sun for two or three days. These leaven cakes were commonly hung up in bags and would keep several months.

Salt-rising, like sugar yeast, did not need any other yeast to start it working. It was prepared from salt, flour, and warm water or new milk, but if it was not carefully made it imparted an unpleasant flavour to the bread.

In towns many inhabitants preferred using brewers' or distillers' yeast, but its quality was not uniformly good. Such yeast sold for 1½ d. a half pint, but as it was generally very bitter an equal quantity of warm water was often added to draw away some of the strength. This did not mix but was later poured off, and a little warm water and flour added to the yeast. Cured in this way the yeast was satisfactory for bread, though it did not keep as well as hop-rising. Sometimes grocers prepared yeast for sale in their shops, but as the years passed, most of these gave way before nationally-known brands, and it became unnecessary and unusual for people to make their own.

(3) BAKING BREAD

Bake-ovens were of several types. In Lower Canada an oven of stone or clay, erected on a stone foundation and having a steep roof of boards, was usually to be found a few yards from the kitchen door. A fire was built in the oven, and when it was thoroughly heated the coals were drawn out and the bread put in. Sometimes a small fee was paid for the use of the *seigneur's* oven, but most *habitants* had their own or there was one oven for the use of a group of neighbours. The loaves were usually of a weight of five or six pounds.

In Upper Canada the bake-oven was more frequently inside the house, at one side of the fireplace. A small door allowed the pans of bread to be placed in the oven. But such a luxury as a fireplace with a bake-oven was not common in a settler's first years. Instead the bake-oven was erected on a large stump, and over the stone or clay sides was placed a roof of bark or slabs. Hollowed buttonwood trees were also frequently used as ovens.

Many of the Loyalists used bake-kettles before they had a bake-oven. The kettle had an iron lid, and hot coals above and below baked the bread. Experience was necessary to produce good bread in the iron pots, for in many instances, while there was a brown crust on the outside, the interior of the loaf was still unbaked dough due to the presence of moisture which could not escape. An improvement was the reflector, an oblong box of bright tin enclosed on five sides of the six. The open side faced the fire when the reflector was laid on the hearth, the tins of dough inside it were raised slightly so that the heat would circulate, and the upper part of the reflector could be adjusted so that the housewife could inspect the contents. In some regions a cylindrical oven called a 'roasting kitchen' was in common use.

Bread as baked by pioneer settlers varied from indigestible 'dough cakes' to first-class loaves. John Langton's first bread consisted of 'round thinnish cakes of good dough put into the bottom of a frying pan', and he found these half-toasted and half-baked cakes quite palatable when mixed with pork fat. Other settlers baked a similar bread in pans with hot ashes above and below, or in improvised ovens of hot stones. This type of bread received no kneading, nor was there any rising through the use of barm or yeast. In Australia such unleavened, heavy breadcakes were called dampers, and some settlers who had heard of their use imitated them in Canada. Early settlers in Peterborough County kneaded and baked their bread in the following primitive manner:

A portion of a trunk of a basswood tree about three feet long and two feet in diameter was split in two halves through the centre. One of these was hollowed out as smoothly as possible, to be used as a kneading-trough. About three pounds of flour, with enough water to wet it thoroughly, was put into this and well kneaded. It was then flattened out and placed in a round, long-handled pan, the front of which was held before the fire by means of a string attached to the end of the handle, while live coals were placed beneath and behind it.

But in wet weather the same settlers found it preferable to place lumps of dough right in the ashes and hot coals. If entirely covered, they baked without burning.

A good bread could be made without yeast of any kind. Mrs. Child's *The Frugal Housewife* gives a recipe which was used by many early settlers who had no yeast and did not know how to make it:

Scald about two handfuls of Indian-meal, into which put a teaspoonful of salt and as much cold water as will reduce the mixture of meal to blood-heat; then stir in wheaten flour till it is as thick as hasty-pudding and set it before the fire to rise. In about half an hour it generally begins to thin and look watery on the top. Sprinkle in a little more flour, and mind and keep the pot turned from time to time, taking care not to let it be too near the fire or it will bake at the sides before it is risen. In about four hours it will rise and ferment, as if you had set it with hop-yeast; when it is light enough mix in as much flour as will make it into a soft dough; grease a pan, put in your loaf, and let it rise, covering it up warm and turning it so that the heat affects it equally; in less than an hour it will be ready for the oven; bake as soon as it is risen. Some bake in a Dutch-oven before the fire.

Satisfactory recipes for bread and innumerable types of cake and biscuits were speedily developed by the women as soon as flour, yeast, and the other ingredients were available. In many instances Indian meal (cornmeal) was mixed with wheat flour, while wheat bread varied in colour and texture from fine white to coarse dark brown according to the proportion of bran remaining in the flour. Mashed potatoes were also mixed with wheat flour by many settlers. Mrs. Catharine Traill gives the following recipe in her *Female Emigrant's Guide*:

Wash and pare half a pail of potatoes, taking care to remove all dark specks; throw them into a vessel of clean water as you pare them, as they are apt to acquire a brownish colour which spoils the white and delicate appearance of the bread. Boil the potatoes till reduced to a pulp, bruising any lumps smooth with a wooden beetle or pounder; it will then have the consistency of thick gruel; when cool enough to bear your hands in it, stir in as much flour as will make the mixture the thickness of thick batter; add a good handful of salt and two cupfuls of your hop barm or any good rising that you may have. A deep, red earthen pot, or a wooden pail, will be a good vessel to contain your sponge. It is a wise precaution to stand your vessel in a pan, as it is apt to flow over. If set to rise over-night it will be risen time enough to work up in the morning early; in summer we seldom make this potato-bread, on account of the potatoes then not being so fit for the purpose, for while young they will not boil down so smoothly; but from the month of August till May it may be made with great advantage. The quantity of sponge above will raise two large milk-dishes of flour, or about twenty pounds of flour. If you have a large kneading-trough you can mix the whole at once and knead it well and thoroughly; but if your trough be too small for convenience divide your sponge and make two masses of dough, working it very stiff on your board, scoring the top with a knife, and cover it up by the fire with a clean cloth; or you may make only half the quantity, using, of course, less potatoes and water. In about two hours, or may-be longer, you will have a light dough like a honey-comb to make into loaves. When baked take your bread out of the pan, wet the crust of your loaves over with clean water or milk, and wrap them in a clean cloth, setting them up on one side against a shelf till cold. This plan keeps the bread from becoming hard and dry. For lightness, sweetness, and economy this is the best bread I know, resembling really-good baker's bread in texture and look. I cordially recommend it to the attention of the Canadian housewife.

(4) PREPARING WILD RICE

Wild or Indian rice grows in large beds in quiet waters of a depth of from three to eight feet. Many shallow lakes with mud or sand bottom are almost filled with the rice plant, and to paddle even a canoe through them in July or August is not easy. As the rice reaches the surface its leaves lie for a time on the water, but as it gains strength it shoots several feet above, and in August bursts into flower.

September is the season of the rice harvest. As the plant withers, the grain ripens, and it is collected in a canoe, the heads being bent over the side of the boat by the paddle while

An early postcard

INDIANS GATHERING WILD RICE

a light stick in the other hand of the paddler strikes the grains into the boat. Among the Indians the squaws usually collected the rice.

When several bushels of rice had been gathered, the Indian built a hedge-like enclosure of pine or cedar branches.

In the centre of this place [writes Catharine Traill] they drive in forked sticks in a square of several feet, across which they lay others, and on this rude frame they extend mats of bass or cedar for the manufacture of which the Indian women are renowned. They light a fire beneath this frame, and when reduced to hot, glowing coals, the rice is spread on the mats above the fire. The green enclosure is to keep the heat from escaping; the rice is kept stirred and turned with a wooden shovel or paddle, and after it is dried the husk is winnowed from it in large open baskets shaken in the wind. This is the mere drying process of the green rice.

The parched rice was then heated in pots over a slow fire, when it burst and showed the white floury part beneath the dark skin. It was then ready for use, and the Indians stored it in birch-bark baskets or sewed it up in mats. They frequently carried a supply with them on hunting expeditions, for a handful could be eaten dry from time to time, without a stop to prepare a meal. Settlers seldom harvested wild rice, for imported rice could be obtained cheap. Many of them purchased a supply from Indians, and found that when boiled or baked it was of a grayish-black colour with a rich and pleasing flavour.*

(5) MAKING STARCH

It was almost a maxim among early Canadian settlers never to buy what could be made in the home. The careful housekeeper consequently made her own starch. Corn starch, used in puddings and custards, was an American preparation which had to be purchased, but starch for clothes could be readily made. There were two main kinds, potato and bran.

In making potato starch white-skinned potatoes were grated on a rasp and stirred in a pan of cold water. The mass was then strained through a cloth, the pulp being squeezed until quite dry. When the liquid settled, a sediment was left at the bottom; and a second and third water replaced the first, each being in turn drained off. The starch at the bottom had then a slightly discoloured crust, but when this was removed the pure

---

*Roast venison, partridge, or wild duck with wild cranberry jelly, and a dessert of wild rice, maple sugar and wild grape juice is a pioneer meal worth eating.

white starch was left. Before being placed in paper bags it was thoroughly dried in the sun. In addition to its general use for starching clothes, this starch, boiled in milk, formed a substitute for arrow-root for invalids and children.

Bran starch was made similarly. The bran was steeped in water in the sun or a warm room until it began to ferment, which was apparent by the bubbles. In a week or two it soured, and after straining off the liquid the housewife squeezed the bran through coarse canvas. The liquor settled, was drained off, and two or three further waters were placed on it in turn, when, as in the case of potato starch, the discoloured starch on top was removed and the clear product left. If it was desired to blue the starch this was done by blueing the last water and stirring well, or else blueing the water in which the starch was eventually used. Sometimes the sour fermented water was used to brighten red and scarlet dyes, while the refuse bran was fed to cattle or hogs.

(6) DISTILLING AND BREWING

The distilling of whiskey was not a prominent domestic manufacture among Canadian pioneers, but was usually left to those who specialized in the business. There were frequently

John Ross Robertson Collection   R. Baigent

JOSEPH BLOOR'S BREWERY, ROSEDALE, 1865

The building was near the new Sherbourne Street bridge. On the hill at the right is one of Toronto's ancient blockhouse forts. The artist was Drawing Master at Upper Canada College.

several distilleries in a small settlement, and it was not unusual
to operate them in conjunction with grist mills, for frosted or
rusted grain and other poorer grades of wheat, corn, or rye,
could be used in the manufacture of whiskey. In fact an early
traveller compared the liquor to 'fire and brimstone', for he
found that the worst of it was 'made of frosty potatoes, hemlock, pumpkins, and black mouldy rye'.

Early settlers could obtain whiskey from the distiller under
much the same terms as cloth from the woollen mill or lumber
from the sawmill. Joseph Pickering says that about three and
a half gallons were made from a bushel of grain, and wrote in
his diary: 'Sent 200 bushels of wheat to the "still", to have
seven quarts of whiskey per bushel for it.'

John Scott of Brampton was locally famous because the
mill-stones in his distillery operated vertically instead of horizontally. There was always a great deal of waste material from
distilleries, and it was used to feed hogs. Philemon Wright,
founder of Hull, notes that he had a shed 500 feet long, with
'troughs to receive the wash for the benefit of the cattle, hogs,
etc.'

The brewing of beer from malted barley, hops, and water
was similarly not a domestic manufacture in pioneer Canada.
There were, however, home-brewed beers for the manufacture
of which extensive equipment was not needed. Hops were extensively grown by many early farmers. 'Every one cultivates
this plant,' says Mrs. Traill, 'on account of it being an indispensable ingredient in making barm for rising the household
bread, besides shading and adorning their verandahs by its
luxuriant foliage and graceful flowers.' She describes how various drinks were made from hops, beets, corn syrup, maple or
birch sap, apples, and yeast, with the addition sometimes of
spruce sprigs or ginger.

(1) FISH RECIPES

FISH were so plentiful in pioneer Canada that it is not surprising that there were numerous methods of cooking them. Besides the more usual frying, boiling, and steaming of fish, there were many recipes not now common. Sometimes soup was made by boiling small fish until they broke into pieces, and then straining the liquor through a colander. This was then made into soup by the addition of flour, bread crumbs, parsley, savory, onions, etc., with a little butter and tomato catsup and some cream or milk. Sometimes other fish, as well as the roes, were added while the soup was being prepared.

Bass or maskinonge were made into fish pie by boiling, removing all bones, and pounding the flesh fine. To it was added a pint of cream, a little butter rolled in flour, chopped parsley, vinegar, the yolks of two eggs beaten, and a tablespoonful of walnut, mushroom, or tomato catsup. Half an hour's baking produced an appetizing dish.

Various kinds of fish were potted after boiling. Between each layer was sprinkled salt, pepper, cloves, and allspice, and when the pot was full, the whole was covered with vinegar and baked in a slow oven overnight, or at least for several hours, a crust of dough serving to keep in the steam. After standing for several days, it made an excellent breakfast or supper dish.

Sometimes maskinonge or salmon were dried in the sun, salt being sprinkled over the flesh, which was turned each day. After two days, the salt was washed off and the fish was strung on a willow wand, hung up in the sun for several days, and then smoked. Red cedar bark and corn cobs were often used to provide the smudge, for they added greatly to the flavour. To prepare for the table the fish was soaked in warm water and boiled or fried.

Stereopticon view, courtesy Clara Benson

FISHING PARTIES, BURLEIGH FALLS, 1880

Courtesy Mrs. R. B. McWilliams   Thomas Workman

PERRY'S CREEK, 1895

Early called the Lost Channel, Perry's Creek is a secondary outlet
of the waters of Lovesick Lake into Stoney Lake.

Canadian eels were of large size, and though the flesh was coarse, it was frequently eaten. Mrs. Catharine Traill recommended parboiling them first. The head, the tail, the backbone, and the oily fat along it were then removed. The fish was spread open, strewed with chopped parsley, savory, pepper, and salt, and rolled into a bolster. Bound and covered with a cloth, it was boiled in salt water for several hours. Mrs. Traill poured vinegar over it when cold, and served it in slices, garnished with parsley.

In winter, fish were usually speared through the ice. If a supply was to be kept through warm weather it was necessary to use salt as a preservative. John Langton used alternate layers of potatoes and fish, placing a layer of mashed boiled potatoes with plenty of salt and pepper in the bottom of a barrel. The fish—maskinonge in this instance—were then boiled, and at a certain stage all the flesh would fall from the bones if the fish was held up by the tail. Salt and pepper were again added, and the process repeated until the barrel was full of potatoes and fish. It was then headed up and left to freeze solid, and the contents would keep well in a cool place until early summer. When he desired it for food Langton fried this mixture in small round cakes. An ice-house made it possible to keep barrels of fish in a very satisfactory manner.

(2) DRESSING GAME

In pioneer days those settlers who did not shoot their own deer could readily obtain the meat from Indian hunters in exchange for salt pork, flour, or farm produce. The dressing of venison and other game did not vary greatly from the preparation of domestic animals. Venison was roasted, fried, and made into fricassees and pot-pies. Sometimes it was corned by rubbing with salt, or the hams were hung for three weeks in the smoke-house after much rubbing with a mixture of sugar, salt, and saltpetre. Venison soup was thickened with wild rice, while jerked venison was prepared by cutting the flesh in strips and drying in the open air.

Lumbermen and hunters often ate the flesh of muskrats, bears, porcupines, beavers, and even woodchucks (ground hogs), but settlers preferred black squirrels or Canadian hare. Black squirrels were considered very delicate food, while bear meat was like coarse beef, though the hams, if well cured,

Eva Brook Donly Historical Museum   W. E. Cantelon

NORFOLK COUNTY COURTHOUSE, SIMCOE, 1837-1863

The painting was based upon an early ambrotype.

Edgar Cantelon could hardly get half a dollar for a painting in his lifetime, but Norfolk County's history is now greatly enriched by some 300 results of his enthusiasm and skill.

Eva Brook Donly Historical Museum   W. E. Cantelon

WILLIAM POPE'S SHOOTING LODGE, VITTORIA

Many of Pope's water-colours of birds and flowers are in the John Ross Robertson Collection, Toronto Public Library.

The Illustrated Sporting and Dramatic News, 1875   Smith Bennett

SHOOTING WILD PIGEONS

Public Archives of Canada   James Pattison Cockburn

'NET FOR CATCHING PIGEONS AT ST. ANNES', 1828

were excellent. Hares were roasted, fried, stewed or made into pies, but were not considered as well-flavoured as the English variety.

(3) WILD PIGEONS

Snipe, partridge, woodcock, and quail, ducks, geese, and wild turkeys were all readily obtained in many districts in the pioneer period, but of special interest is the wild pigeon, which was exceedingly plentiful in the spring and fall. Stories of huge flocks literally clouding the sky in the migration seasons are not exaggerations, and many a settler added variety to his food supply by shooting, netting or knocking them down in great numbers. They were found to be good cooked in almost any manner, but were best roasted or baked in pot-pies. The old-fashioned bake-kettle made a better pigeon pie than the stove oven. A good pie-crust lined the kettle, and the birds were placed inside, with a little butter and seasoning on the breast of each. A cupful of water and a half-inch crust over all completed the pie, which was baked by keeping hot embers underneath and on the lid and turning the bake-kettle from time to time.

The pigeons were fattest just after the wheat harvest, and some settlers were accustomed to salting them down in barrels, the breasts only being used. Packed in the ice-house they provided a delicacy for many weeks, while their feathers were used to fill mattresses. Stuffed specimens may be seen in museums, but the wild pigeon is now believed to be extinct.*

*In a letter to the Toronto *Globe and Mail,* February 1, 1940, the Hon. Mr. Justice W. R. Riddell says: 'In my youth it was usual for millions—literally millions—of wild pigeons to pass over my father's farm in the Township of Hamilton in the spring for days at a time. I have more than once seen them break branches of trees by a great number of them lighting at once on the same limb, and I twice killed three by one shot. The last I saw—they had been gradually diminishing for years—were at Hastings in 1872; there I killed two by one shot.'

(1) MAKING SAUERKRAUT

SETTLERS of varying nationalities had usually a fondness for prepared foods which, in their native land, had something of the characteristics of a national dish. The German food, sauerkraut, affords a prominent example. Apart from their customary use of this food in Germany, there was good reason for its use as a winter dish in Canada, for green vegetables, and fresh foods generally, were not obtainable.

Cabbage is the prime ingredient of sauerkraut. Having gathered the cabbages and trimmed off their outer leaves, the farmer proceeded to cut the rest up fine with knives or a spade. A common method was by the use of a home-made cabbage-cutter, with knives projecting slightly above a board. On this board was fastened a box without a bottom, raised above the knives by cleats at the sides. The board was placed over an open barrel, and the box filled with cabbage and pushed back and forth over the knives. As the shredded cabbage fell into the barrel a generous quantity of salt was sprinkled over each layer. The cabbage was pounded down closely, and boards surmounted by heavy weights covered the barrel when full.

After several weeks the sauerkraut was ready to use. When the barrel was opened, brine covered the sauerkraut and kept it from spoiling. Those whose stomachs were not used to strong food were inclined to refer to sauerkraut as 'rotten cabbage', but German settlers ate large quantities of it during the winter months.

(2) APPLE CIDER MAKING

Apple cider was a common drink in early Canada. In many districts, we are told, 'It was a universal custom to set a dish of apples and a pitcher of cider before everyone who came to the house.' Sour apples were used for cider, while sweet apples greatly aided in thickening the boiled syrup when apple sauce, apple butter, preserves, etc., were being made—the great source of apple pie, long a standard dessert.

The pioneer cider mill was a rather crude contrivance. It consisted of two cogged wooden cylinders about fourteen inches in diameter and double that in length; sometimes one was slightly longer than the other, and to its pivot was fastened a pole which a horse pulled while walking around a circular path, revolving the cylinder. The cylinders stood upright in a framework, and as they revolved they crushed the apples and shot them out on the opposite side.

Before the introduction of the screw press a cruder one was in general use. On a heavy beam with an upright at one end was placed a square box of hardwood slats. Another beam, with one end mortised to the upright, extended over the box and carried at its end a weight so arranged that by turning a wooden screw it could be lowered upon the apple pulp in the box.

Sometimes the press was as large as eight feet square, and round it was a deep groove to carry off the juice. In making what is called the cheese a thick layer of rye or wheat straw was spread on the floor of the press around the outer edge, and on this the pulp was spread a foot or more deep, sometimes in a cloth bag. The first layer of straw was then turned in carefully, and the process repeated several times until the press was full. Planks or wooden blocks were placed on top, and by the use of a wooden screw a great pressure could be exerted. Casks or vats caught the juice as it was squeezed out. Some barrels were set aside as rack cider for bees and other festive occasions, and this 'apple Jack' soon became hard and intoxicating. Other barrels of cider were destined to be used to boil down for preserves or apple butter, while cider vinegar was readily formed by exposure to the air for several months.

There were cider mills where those might be supplied who did not make it at home. Apples were plentiful, and the cider sold in 1830 for from 4s. 6d. to 9s. a barrel. In some districts you could take two barrels to the mill and receive one back full for the empty; and in others 1s. a barrel was charged for the use of the press, the farmer supplying his own apples, barrels, and labour.

(3) MAKING VINEGAR

Nowadays the vinegar one buys at the grocer's is usually malt, cider, or white wine, but in the pioneer period, vinegar was generally home-made. The sap of the birch tree, as well as that of the sugar maple, was found to make good vinegar. From three to five pails of sap was boiled down to one, the sap being best at the close of the season when it was poorest for sugar. While it was still blood-warm it was strained and a teacupful of rising or yeast added. The liquid was then set in the sun or near stove or fireplace to ferment. After it had worked for some time and had stood for several weeks it was ready for use.

By exposing apple cider to the air for several months, a good cider vinegar was made. A quick method of making vinegar was occasionally employed. A barrel of apple cider was laid on its side above a barrel of birch shavings. The cider was allowed to drip a drop at a time into the shavings, and its slow progress through them led to its oxidation from the air. From the spigot at the bottom of the barrel, it dripped after many hours—perhaps a whole day—into a third barrel. Incidentally extracting birch flavour from the shavings, the liquid was also immediately converted into vinegar in the process.

Just as cider vinegar was closely related to apple cider, so the recipe for beet vinegar followed that for beet beer, but the fermentation was allowed to go on much longer. This vinegar was noted for its fine colour. Another vinegar was made from red or white currants. They were usually strung, and then placed in an equal volume of water in a warm place to ferment for several days. Stirring from time to time prevented mould accumulating on the surface. To the juice which drained off and could be squeezed from the pulp was added two pounds of coarse sugar per gallon, and the cask was placed in a warm room. In six weeks' time a strong and fine-coloured vinegar was ready for use in pickling.

(4) PREPARATION OF DRUGS AND REMEDIES

Medical knowledge was comparatively primitive in the pioneer period, and qualified doctors were few. As a result there was a dependence upon natural remedies on the one hand, and at the same time upon quacks, who preyed upon the ignorant. Magic, charms, and miracle cures found many believers, especially among the poor, and herbalist healers, both men and women, practised their trade in every rural district.

Mere mention of the remedies for various diseases will indicate the extent of the belief in herbs as cure-alls. An infusion of bark was prescribed in large doses for fever patients. Seneca snake-root was believed to relieve fever, colds, and 'pains in the bones'; and combined with thimbleberry and blackberry roots, it was used to treat cancers. Wild horehound, sweet gale, and the leaves and roots of the strawberry were brewed as a remedy for ague, and burdock roots for indigestion; while alum root and crow's foot were thought to relieve dysentery. The Indian use of the bark of the sassafras or spice-wood tree was sometimes imitated: the chips of bark were boiled and the liquor had a fine aromatic flavour, reputed to be an effective purifier of the blood. Cherry bark was also a regulator of the blood, and the bloodroot or gentian root, dried and ground, was prescribed for rheumatism, and sometimes for certain affections of the nose.

Many considered that herbs must be gathered only when the phase of the moon was favourable. Roots, stems, leaves, berries, and seeds were commonly made into 'tea', and the more bitter the concoction, the more effective it was supposed to be. Wormwood tea, for example, was a general tonic when all else failed. Boneset tea and hemlock tea, made from the tips of the tree, were considered good remedies for colds, and yarrow tea for malaria; spearmint tea, spikenard tea, and tea from May-apple roots were used for colds, coughs, and sore throat; and catnip tea for children's ailments.

Methods and instruments were as primitive as the remedies. Bleeding, or letting blood, was supposed to effect an improvement in almost any disease, while protection from germs—even their existence—was almost unknown. Necessary operations were carried out while the patient was conscious, and often without the benefit of drugs to relieve pain. In rural districts there were even substitutes for surgical dressings. Crushed plantain leaves were considered effective as a poultice, while smartweed steeped in vinegar aided in reducing bruises. Many years were to elapse before expert medical attention was available to rural settlers, and even now it is insufficient in most regions.*

*The most detailed account of early medical conditions is W. Perkins Bull, *From Medicine Man to Medical Man* (1934). It has particular reference to the County of Peel.

(5) SUBSTITUTES FOR TEA AND COFFEE

Besides the various 'teas' which have been described as remedies, there were a number of substitutes for tea and coffee, which many early settlers found too expensive for general use. The leaves of the wild plant New Jersey tea, or mountain sweet, supplied a common substitute for tea. The leaves were cured by drying in the sun or an oven, after which they were rolled. Lumbermen sometimes made use of this plant, as well as another, Labrador tea or marsh rosemary, whose leaves when boiled produced a liquid of strong resinous flavour. Tansy tea was also popular, while maidenhair fern, cherry bark, sage, thyme, and chocolate root provided drinks both pleasing and medicinal.

Raspberry vinegar, a drink popular among early settlers, is still considerably used in Canada. It is not, of course, a vinegar, but was made by letting raspberries stand twenty-four hours in vinegar, draining off the liquor, and repeating the process a second and third time, when the vinegar is sufficiently infused with raspberry juice. A pound of lump sugar was added for each pint of juice, the stone jar of liquor was placed in a pot of boiling water, and after ten minutes' boiling it was bottled. The flavour was best when the bottles had stood in a cool place for several months, and only a little was needed in a cup of water to make a palatable drink. It was found that the use of the poorer grades of sugar made an inferior, though cheaper, drink.

Substitutes for coffee were numerous, ranging from beans, peas, barley, and corn to dried potatoes, rye, wheat, and even toasted bread. Acorns were among the forest products occasionally boiled for 'coffee', while dandelion roots were prepared as follows by Mrs. Susanna Moodie:

I carefully washed the roots quite clean, without depriving them of the fine brown skin which covers them and which contains the aromatic flavour which so nearly resembles coffee that it is difficult to distinguish it from it while roasting. I cut my roots into small pieces the size of a kidney-bean, and roasted them on an iron baking-pan in the stove-oven until they were as brown and crisp as coffee. I then ground and transferred a small cupful of the powder to the coffee-pot, pouring upon it scalding water and boiling it for a few minutes briskly over the fire. The result was beyond my expectations. The coffee proved excellent—far superior to the common coffee we procured at the stores.

(6) APPLE RECIPES

Apples were almost as characteristic of early Canada as they are today, for many of the first settlers planted orchards, and the apple-paring bee was a typical rural gathering. 'Bushels and bushels of apples', wrote a pioneer woman, 'are pared, cored, and strung on Dutch thread by the young men and maidens, and the walls of the kitchen festooned round with the apples, where they hang till dry and shrivelled.' If they were dipped in boiling water before hanging, the colour was preserved. Sometimes they were hung outside the house, in the sun and wind, or spread out on boards or trays, and when thoroughly dry stored away in bags.

Dried apples provided fruit all the year round, for a short stewing rendered them tender. The sour kinds needed a little sugar or molasses, but others were sweet enough as they were. Tarts, pies, apple preserve, and apple jelly were favourite dishes; in fact, apple pie became the standard dessert in town and country. Apple syrup, apple butter, and apple sauce were names for much the same product, and sometimes apple cider was boiled down in making them. Among less common apple recipes was that for apple rice, which was as follows:

Wet a pudding-cloth; place it in a basin or colander, having previously well washed and picked a pound of rice, if your family be large; half the quantity will be sufficient if small; place some of the wetted rice so as to line the cloth in the mould all round, saving a handful to strew on the top; fill the hollow up with cored apples and a bit of lemon peel shred fine, or six cloves; throw on the remainder of the rice; tie the bag, not too tight, as the rice swells much; and boil a full hour, or longer if the pudding be large. Eaten with sugar, this is an excellent and very wholesome dish; acid apples are best, and are so softened by the rice as to need very little sugar to sweeten them.

(7) PRESERVING SMALL FRUITS

Huckleberries, blueberries, raspberries, currants, cherries, bilberries, and other small fruits were often dried for future use. After boiling for half an hour, the fruit was spread out on pans and dried in the oven, under the stove, or outdoors in the sun and wind. When partially dried it was cut into squares and turned over, and as it dried further, white sugar was sprinkled over the fruit and it was pressed tightly and packed away in bags or boxes. These fruit cakes were prepared for use by stewing, which rendered them like preserves. A remedy for a sore throat or cold was made by pouring boiling water on one of the cakes, particularly if it was made of black currants. Many people used dried fruit of this type in place of the currants or raisins which could usually be purchased in the village store. Dried tomato and pumpkin were similarly packed in bags by many farmers.

High- and low-bush cranberries were readily accessible to early settlers, but were scarce even a century ago. The berries were sometimes kept a considerable time by being merely spread out in a dry room. Another method of preserving them consisted of putting them in jars or barrels of cold water. Cranberry sauce, served with venison and other roast meats, and cranberry jelly made from the high-bush fruit, were the chief products used in the home. Among other native fruits occasionally preserved was the mandrake or May-apple, which was boiled in syrup with ginger and cloves.

(8) MAKING GRAPE WINE

In the Niagara Peninsula particularly, the culture of peaches, grapes, and other fruits was important from the earliest years of settlement, but the making of native wines was not prominent

Courtesy of the Old Mill Hotel
CIDER PRESS ON THE HUMBER RIVER IN THE 1870's
This large press was located near the Old Mill.

until more than half a century had passed. At the time of Confederation the industry was being developed on a large scale
at Grimsby, and there was an extensive vineyard and winery
at Cooksville. The best red wine flowed from the presses by its
own weight, while inferior white wine was obtained by pressure.
Arched cellars, pure air, and controlled temperatures were important elements in the fermentation, and the phase of the moon
had much to do with success in the processes. In 1866 and
1867 the Cooksville winery of Messrs. Cooke and de Courtney
produced 90,000 gallons of wine.

A few settlers made a substitute drink from maple or birch
sap, yeast, and sometimes a little sugar and split raisins, but
such wine went sour quickly.

Some pioneers were content with wild grape juice, and
Catharine Traill found that a fine jelly could be made. She gives
the recipe, as well as adding a means of beautifying the land-
scape by transplanting grape vines:

From the wild grape a fine jelly can be made by pressing the juice
from the husks and seeds and boiling it with the proportion of sugar
usual in making currant-jelly, *i.e.*, one pound of sugar to one pint of
juice. An excellent home-made wine can also be manufactured from
these grapes. They are not ripe till the middle of October and should
not be gathered till the frost has softened them; from this circum-
stance, no doubt, the name of Frost-grape has been given to one
species. The wild vine planted at the foot of some dead and unsightly
tree will cover it with its luxuriant growth, and convert that which
would otherwise have been an unseemly object into one of great
ornament. I knew a gentleman who caused a small dead tree to be
cut down and planted near a big oak stump in his garden, round
which a young grape was twining: the vine soon ascended the dead
tree, covering every branch and twig and forming a bower above
the stump, and affording an abundant crop of fruit.

(1) HOG-KILLING

A HEAVY and disagreeable task on the pioneer farm was the slaughter of live stock. Two or three experienced men could kill and dress a dozen hogs in one day, but they were usually helped by the women and children. Among the operations outside the house was the removal of hair and bristles by hot water, while the housewife was soon busy inside getting everything ready for the making of sausages, lard, and other meat products.

(2) SALTING MEAT IN BARRELS

Within a few hours the carcass was cut up and stored away. The hams were first cut out, and then the forelegs in ham shape; the rest was cut in pieces chine fashion, and all the parts were rubbed and packed as tight as possible in salt in tubs, casks, or barrels, which were then filled up with strong brine and headed. Altogether the farmer prepared and stored a large supply of fine meats for his family's use, and probably several barrels of hams and other cuts for sale, for salt pork was in steady demand for the British garrisons in Canadian forts. Two hundred pounds was the usual weight of a barrel of pork. When it contained only the side pieces it was called mess pork; if hams and shoulders as well, prime mess, while all parts of the hog packed indiscriminately went by the name of prime pork. Pigs were often sent to market in a frozen state, without cutting up, and dried or smoked hams were frequently sold separately by the pound.

(3) PICKLING IN· BRINE

Some people preferred packing their meat in a brine solution rather than dry-salting it. One recipe called for the following

Conservation Authorities, Ontario   Snider Collection

PRESS-FILLING MACHINE AND MEAT GRINDER

ingredients for pickling 200 lbs. of meat: 14 lbs. salt, ½ lb. saltpetre, 4 lbs. coarse brown sugar or 2 qts. molasses, 1 pt. of beer or vinegar, and enough water to dissolve the salt. Pepper, allspice, and cloves were often added to give flavour to the meat. The liquor was brought to a boil and the scum removed. When cold it was poured over the hams, which had meanwhile been well rubbed. The meat remained in this pickle for six or eight weeks, being turned and basted every two or three days. The hams were then ready for the smoke-house. There were, of course, many variations in the brine; often a little alum or potash was added. Mrs. Traill gives the following directions for preparing bacon for smoking:

Having taken off the hams from a side of pork, chop the rib-bones close to the back so as to remove the backbone the entire length of the side. With a sharp knife raise all the small long bones from the meat, and trim all rugged portions carefully away. Then mix a pound of coarse sugar to 2 oz. of saltpetre and 4 lb. of salt. Rub this well over the meat on all sides; two sides of bacon will not be too much for the above quantity. Cut them in two pieces and lay each piece above the other, the rind downward, and strew the remainder of the salt mixture over the last piece. A shallow wooden-trough or tray, with a hole and peg at the bottom, is the best to salt your bacon in; it should be placed a little sloping forward. Every second day draw off the liquor that runs from the meat into a vessel and carefully pour it over the meat again, having first shifted the bottom pieces to the top. In six weeks' time take them out; rub with bran, and lay on the rack to dry, or smoke them; this process makes excellent meat.

(4) MAKING LARD AND SAUSAGES

While the men were engaged in cutting and packing, the women were removing fat and trying (melting) it into lard. The kidney-fat of the hog was cut up small and boiled down over a slow fire. As it boiled, the oil was extracted, but when the process was almost complete the danger of burning was greatest. The oil was then strained off into stone jars and covered. Suet was rendered at the same time.

The liver was also cooked and liver-wurst made, while head-cheese and pickled souse were prepared after soaking the meat around the head and feet and chopping it fine.

Sausage-making was another process frequently completed the same day. Parts of the entrails provided a covering for chopped and flavoured meat, for early sausages were more like meat cakes than the modern sausage. Cotton bags were also used as casing to hold the meat.

(5) MAKING GELATINE

Many early Canadian settlers made their own gelatine, which was formed by treating ground bone and cartilage with boiling water. When heated, gelatine is a liquid, but when cool it is a jelly, so it is used to harden meat dishes, desserts, etc. It is made from much the same materials and in much the same fashion as glue, but with greater care in the selection and cleaning of the ingredients. As the slaughtering of cattle became a specialized industry gelatine became one of its natural by-products.

(6) SMOKING MEAT

Many settlers erected a smoke-house to enable them to cure their own meat in the spring. The hams and beef which had been salted down in casks or tubs were removed from the brine, washed, and hung in the smoke-house, which was a wooden structure. If the farmer had an outside bake-oven the smoke-house was near by. Its wooden walls were covered with grease, and the smudge for smoking was best obtained by burning beech, birch, hickory, maple, or corn cobs. Beechwood creosote and other oils cured, preserved, and flavoured the meat. In later years many farmers had improved smoke-houses of brick or stone. Sometimes the meat was smoked for only a few days, but others made a practice of keeping it in the smoke-house for six weeks or more. The hams, sewn in linen or cotton bags, would then keep for months, even through summer heat.

A coating of whitewash over the cloth covering of the bags aided in keeping flies from the meat, but weevils sometimes penetrated and laid their eggs in the skin and joints. As a result, hairy worms were soon destroying the meat. Immersion in boiling water, rubbing with bran or sawdust, and packing in wood ashes or oats usually restored the meat and permanently rid it of the weevils. As a preservative, charcoal was often packed in the barrels with the hams, and it was also used to restore tainted meat; a pint of the drippings from the stove-pipe joints was sometimes added to the brine for the same purpose. Cedar bark burned in barrels acted as a purifier when the brine had stood in them too long.

(7) CANDLE-MAKING

Although candles were not difficult to make, they were often scarce, especially among 'backwoods' settlers.

One was lighted for supper [says a farmer] but it was put out immediately after the meal; and we had to sit at the light of the fire, which we made as bright as possible by a supply of resinous pine. . . . Some lard in a saucer, with a piece of rag for a wick, was one of our plans, in addition to the pine, when we wished to see our way to our beds.

In some districts this substitute of twisted rag was called a *witch*. At Johnson's Inn, near the present city of Barrie, a traveller in 1823 was informed by the landlady that she had no candles. But she quickly ran a thick cotton thread down a candle-mould, filled it with melted lard, sank it deep in the lake for fifteen minutes, and provided for his use what she called a 'country-made lamp'.

Many settlers used torches of pine knots or rushes, particularly to light their way through the woods. On that account fat pine was called 'candle wood', and was stored up for winter use. The knots gave a fair light, but tar dripped from them and made the air smoky. Similarly the dish of fat had its disadvantages, for the light was poor, the odour unpleasant, and soot was scattered about as a result of its use.

There were two general types of candles—tallow dips and mould candles. The use of moulds was less arduous work and produced more regular candles, but many found the dips more durable. The season for candle-making was autumn, when tallow was rendered down after the killing of beef and mutton.

Courtesy of the Langton family   Anne Langton

CANDLE-MAKING MACHINE

Often the tallow, or hard fat, was saved over a period of several months, so that a large batch could be made at one time. Tallow was sometimes melted in a large pot of boiling water, in the proportion of one-third tallow and two-thirds water. The tallow rose to the top and was well strained through flannel over a colander. Other settlers used only a tablespoon or so of water to prevent the fat burning to the bottom of the pot. Small pieces of fat were thrown in and boiled over a slow fire until all the fat was 'tried down'. It was stirred from time to time with a stick, for a metal spoon soon got too hot to hold. The scraps left after the tallow was strained were not wasted but set aside for soap grease.

The pot of tallow was placed on the floor between two chairs, a foot and a half apart. Two long poles were laid over the backs of the chairs, and on them were placed every few inches the candle rods, which were about twenty inches long and half an inch in diameter; on each of these, six pieces of coarse cotton wick some nine or ten inches long were placed at equal intervals, and a year's supply of candles might be made on fifty rods.

Two kettles were often used for the tallow, one being on the stove all the time so that the temperature was just right for the dipping, which consisted of putting each rod with its six wicks into the liquid tallow for the right number of seconds and placing it back on the poles. When all the wicks had been dipped the process began again, for many dippings were needed to make the candles the required size. If they remained too long in the tallow, or the time between dippings was not right, the candles would crack and be otherwise unsatisfactory. With two kettles one man could make two hundred candles in a day. Cool air was essential to harden the tallow between dippings, and the small boys of the family were often given the unpleasant work of holding the dips outside in the cold while other workers were dipping inside.

The better candle-makers used moulds. These were suspended from a small frame, and the wick was drawn through the bottom of the tube and threaded through the small pointed end which formed the top of the candle. To thread the wick well was a skill in itself. A double length of wick, with a little over for tying, was measured off. Several slender sticks like skewers were in readiness to support the wick, which was slipped through the bottom of the mould loop uppermost. When all the wicks were run, the sticks were slipped through the loops at the top and made even, and then the ends of the wick at the bottom of the mould were tied tightly. The sticks had to be perfectly straight or the wick would not be in the middle of the candle. Sometimes a bar was used above the mould in place of the sticks, all the wicks being fastened tightly to it.

A metal jug with a spout was best used to pour the hot tallow into the moulds. As the tallow cooled it shrunk, and more was poured in to fill up the mould. Since they had to harden all at once, instead of in layers as in the case of the dips, mould candles were not always as satisfactory. But if the tallow was of good quality the knot was cut off at the bottom and the candle readily drawn out of the mould by the stick. Sometimes it was necessary to warm the moulds slightly, or even to pour boiling water around them to release the candles. They were immediately packed away in a cool place where mice could not get at them.

The snuff, or charred wick of the candle, was generally removed by the thumb and forefinger. Sometimes this excrescence was called the *thief,* and the more careless among the menfolk merely threw it on the floor. But the greater gentility of housewives demanded something better, so the candle-snuffer was shortly introduced. With this was a snuffer-box for the charred wick, and even the farmer who had been continually scolded for not making use of it was constrained to admit: 'That's a dandy good rig!'

The economical housewife melted down and strained all drippings or ends of candles, or else threw them in with the soap grease. Lard alone was too soft to make candles, but it was sometimes mixed one part to three with beef or mutton suet. It was possible to make a clear and firm candle from melted lard by the addition of one ounce of nitric acid and a

John Ross Robertson Collection

FREELAND'S SOAP AND CANDLE FACTORY, TORONTO
Located on the wharf at the foot of Yonge Street, 1832-65.

little beeswax to each eight pounds of lard, but very few settlers knew of this recipe, nor were chemicals generally accessible. Some candle-makers steeped the wick in a strong solution of saltpetre before putting it in the moulds, and this was found to brighten the flame and lessen the odour.

In later years it became usual for men to make a business of candle-making, and they would go from house to house with candle-moulds and make the winter's supply from tallow supplied by the farmer. In villages and towns there was usually at least one chandler, and candles were made and sold at his factory. Near the Atlantic coast some settlers had spermaceti (whale-oil) candles, which were much better than tallow. Sperm-oil lamps were in use in the homes of the well-to-do in Canadian towns. Handsome globe lamps burning melted lard were a substitute which could be used at half the expense, but they were not easy to light or keep clean. Tinsmiths sold portable tin lamps using lard for 1s. 6d., but candles were not rendered out-of-date until the last quarter of the nineteenth century, when coal-oil lamps of pewter or glass, which were introduced in the 'fifties, became common. A few large towns had gas plants in the eighteen-forties and -fifties, but not until the 'eighties and 'nineties did electric arc lights illuminate streets, public buildings, and large stores. Today candles have only an occasional, but often a decorative, use in Canada.

(1) BUTTER-MAKING

THE quality of butter depends on the feed of the cattle, the proper treatment of the milk, and the skill of the butter-maker. Many pioneer settlers had defective feed for their cattle and little or no equipment for the preparation of dairy products; in fact, milk was often scarce, even among farmers, and lard and grease frequently served as butter. Cleanliness and fresh air are essentials in dairies at all times, while coolness in summer and warmth in winter are almost as indispensable. The best dairies of the pioneer period had stone foundations, and sometimes the walls were made of upright squared cedar posts, with latticed and shuttered windows. Frequently they were built near a running stream or spring.

Pans of thick glass were in general use, while there were others lined with zinc or 'iron stone', a type of enamel. Wooden trays four inches deep, with peg holes for letting off the milk beneath, were early used, but the difficulty of keeping them perfectly clean led to their being replaced by other types. Earthenware or stone jars were generally considered the best for storing cream.

Churns were of great variety, and many a primitive make-shift served the purpose. Four short planks nailed together made the first churn on one farm, and the housewife said that she made 'as good butter in that churn as any I ever made in my life, but I needed to watch the seams carefully'. The principle was the same in all churns, the cream being pounded and mixed by a plunger until butter was formed. One of the older types was the upright churn, worked with the staff and cross-dash or paddle, but a great deal of labour was needed to operate it. The box churn with sloped sides left no sharp angles and corners to keep clean; the sides had dashers, and another was attached to the beam of the handle, which could be unscrewed when the buttermilk was drawn off through a plug-hole in the bottom. Some churns were worked with an iron wheel turned by a winch. The barrel-churn, with its rotary dasher, was simple and effective, and enabled the washing of butter before its removal. Whatever the type of churn, the scalding of the cream* just before churning was found advantageous, though a common alternative was pouring hot water into the cream, which made the butter white, greasy, and generally inferior. Frozen cream made frothy butter, or none at all; while in hot weather the churn was often cooled by immersion in cold water, either before or during the churning.

When the butter had been churned, it was well washed with cold water to remove as much of the buttermilk as possible, and it was then salted to suit the taste; during the process it was well worked by hand. The butter was then packed in crocks or stone jars, some makers adding two and one-half pounds of salt, six ounces of saltpetre, and a half pound of fine sugar to each thirty-two pounds of butter; and brine having been poured over it to a depth of two inches, the cover was pressed down tightly over a white cloth. So packed, the butter would keep for two years. When the cows had access to turnips or to wild leeks, a little saltpetre dissolved in the cream removed some of the objectionable flavour in the butter, though many settlers used such milk only to feed their calves. Many a storekeeper had to throw out 'turnipy' butter taken in trade in a day when merchants commonly took all of a farmer's butter and eggs.

While cheese was early made in co-operative factories in Upper Canada, it is only in comparatively modern times that *creameries* have partially superseded the individual butter-maker.

(2) CHEESEMAKING

Early settlers frequently made their own cheese, though the process was a laborious one. Making the rennet was the first step. The maw or first stomach of a sucking calf was always used for this purpose, though the stomachs of other milk-consuming animals, such as the lamb, kid, or sucking pig, will also curdle milk. Emptied of all curd and slime, the maw is carefully

---

*Scald cream used to be common enough at farmers' markets, but unfortunately it is now a rare delicacy.

turned, washed, and placed in a brine of salt and cold water for twelve hours. It is then rubbed with salt and stretched on a flexible stick by bending it and holding both ends in one hand. Over this the bag is drawn and tied at the open end, near the ends of the stick, whereupon it is hung up in the sun or in the house to dry and harden. Then the stick is taken out, and the rennet bag is hung up in a cool place in a paper bag; some people placed it in a crock of brine, closely covered, while others filled it with salt before hanging it up.

The rennet might be used in a few weeks, but some considered it best to keep it a year. Even those who did not make cheese preserved the rennet and used it to make sweet curds and whey to be eaten at dinner or supper with fresh fruit. A dessert spoonful of the rennet would set a dishful of milk of the proper temperature, and pioneer children were as fond of it as those of modern times are of junket, which is set by a powder corresponding to the rennet.

A solution of the rennet, perhaps half a cupful for a good-sized domestic cheese, was stirred in to curdle the milk. In twenty minutes or half an hour the curd was formed and separated from the whey. With a saucer, wooden dish, or wooden cheeseknife, the curd was cut across in several directions and the whey rose between the gashes. Usually a cheese basket— a loose square or round receptacle without a handle—was used to break up the curds formed by the souring and curdling of the milk. It was set on a frame called a cheese ladder, which was placed across the tubful of curds. A thin cloth was laid in the basket, with the edges hanging over the sides, and as the curd was ladled into the basket, the cloth was gently pressed to squeeze out the whey, which was drained off, leaving only the curds.

The curds were then put back in the vat, usually in a cloth, and cut up fine. A little salt was mixed through for seasoning, the cloth was folded over, and the cheese placed in a press under low pressure at first. In sixteen hours the cheese was sometimes removed to a tray and salted again, or it was laid in brine for several hours. A linen cloth to bind it together and prevent bulging or breaking was usually placed around the cheese at this time.

The first cheese presses were generally clumsy, with a long lever and great stone weights which had to be lifted frequently as the cheese was periodically rubbed during the drying and curing. Sometimes a small self-pressing cheese mould was used in the home. It consisted of a thin cylinder about a foot deep and eight inches in diameter. All over its surface, at intervals of two inches, it was perforated with holes, and a movable lid fitted on either end. After the curd was fully set it was put in the cylinder, any remaining whey squeezing readily through the holes without other pressure. Every hour or two for a couple of days the mould was turned upside down, when the cheese was firm enough to turn out on the salting tray. Sometimes a wide wooden hoop was then put around it until it was time to remove it to the shelf for drying. Linen binders were used to keep it in shape, and perhaps a coat of whitewash was brushed on the rind to keep flies away.

To make cheese on a large scale was at first very heavy work, with little in the way of mechanical appliances to lessen the lifting. Screw presses were later introduced, and cheese ladders held the large cheeses as they were turned, rubbed, and salted, for the flavour became increasingly rich as the cheese ripened under careful supervision.

Housewives developed methods of cheese-making which suited a small supply of milk and a somewhat makeshift equipment. 'Bush cheese' was a type made by combining the night's milk with the morning's, bringing the whole supply to the heat of new milk; then the rennet was stirred in and the curds drained from the whey and wrapped in several thicknesses of cloth to prevent drying out. The next day's milk was similarly treated, and if the farmer had only one or two cows, perhaps the following day's as well. Then all the curd was put together and well mixed, salted slightly, and pressed. Some settlers made their first cheese with no other press than a board and heavy stones.

A richer cheese resulted from the mixing of the milk and cream of the night before with the new morning's milk, in imitation of the method used in Cheshire. A little flour sprinkled between the layers of curd made a still finer cheese, known as 'blue moulded'. Sometimes a pinch of saffron, steeped in warm water, gave a richer colour to cheese. 'Potato cheese' was made by mixing mashed potatoes with the curd. A pioneer cream cheese was made by the following recipe:

Take one quart of rich cream, when well soured; put it in a linen cloth and tie it as close as you can, as you would a batter-pudding; hang it upon a hook, with a pan below it, to drain for two days; then turn it into another clean cloth and let it drain for another two

TAKING UP THE MILK.

AT THE FACTORY.

RECEIVING & WEIGHING.

IN THE VAT.

THE CURD IN THE SINK.

THE PRESS ROOM.

CURING ROOM.

WEIGHING, MARKING & BOXING FOR MARKET.

Canadian Illustrated News

J. C. Dyer

CHEESE-MAKING IN ONTARIO, 1876

days, till it becomes solid; then lay it on a clean fine cloth spread on a plate; fold the cloth neatly over on each side and turn it over in the cloth on the plate; lay another smaller plate over it, turning every six hours; sprinkle a little fine pounded salt, and lay vine leaves over and under to ripen it; it is fit to eat in a few days, when slightly coated.

German settlers made other types of cheese from their sour milk. Schmier kase is made from scalded thick sour milk. The heat separates the curds from the whey, and a sour curd cheese was the result when the whey drained out of a cloth bag. Cream was sometimes added when the cheese was ready to eat. Hand kase, or ball cheese, was prepared similarly, but was seasoned with salt and butter and rolled by hand into balls before being laid away to ripen. Pot cheese was seasoned curds packed away in a crock in a warm place, and was noted for its strong odour.

Cheese is best made on a large scale, and cheese factories, to which the farmers of a district brought their milk, were among the earliest to develop in place of home manufacture. Most of them were co-operative in management, and profits were apportioned among the members. A typical early factory

Sigmund Samuel Canadiana Gallery   Major-General James P. Cockburn

'QUEBEC FROM POINT LEVI'
These volumes are enriched by numerous paintings by James Pattison Cockburn. 'Colonel Cockburn', wrote Thomas Hamilton in 1832, 'is an accomplished artist, with a delicate perception and fine feeling of the beauties of nature.' He is shown sketching in this picture. The Public Archives, Ottawa, lists him as 'Sir James', but he was never knighted.

stood on a hillside in Cayuga in 1867. The basement was entirely open on one side, and there the cheese was made, after which it was cured in the floors above. The farmers brought their milk morning and night, and after its weight was recorded it was carried by conductors to the vats, where, if it was the evening milk, it was kept cool by spring water continually flowing in a metal chamber under and around it.

In the morning a fire in a furnace beneath brought the water to a temperature of 85°, when the rennet was put in and the milk allowed to stand forty minutes. The mass was then considerably hardened and was cut up, standing some fifteen minutes under a temperature of 96°. The whey rose and was drawn off by a syphon, and the curd was constantly stirred while it was being scalded. The curd then settled and was dipped into a cooler, which was wheeled into the press-room after the curd was thoroughly cooled, washed, drained, and salted, about three pounds of salt being used for each 1000 pounds of milk.

The curd was then put into the hoops, under pressure of a screw, and after half an hour the cheeses were taken out and turned. When they had stood in the presses for twenty hours longer they were conveyed to the drying and curing room, where they were continually rubbed and turned. From twenty to thirty cheeses daily were prepared over a period of some eight months each year, and the cheeses, weighing about sixty-five pounds each, were ready for market in from thirty to fifty days.

Modern cheese factories use steam to heat the milk and to turn the large paddles which mix the curds and whey* for about two hours. Colouring matter, 1½ ozs. to 1000 lbs. of milk, is usually added to the milk before the rennet. The rennet solution is now made in chemical laboratories, sometimes being imported from Germany. The proportion of rennet changes with the season, type of milk, etc., varying from 2 ozs. to 4 ozs. per 1000 lbs. In general it takes twelve pounds of milk to make a pound of cheese. It may be used ten days after making but is known as new cheese for several months. Cheese becomes richer as it ages, and consequently brings a higher price; sometimes it is stored for from seven to ten years.

*The old English nursery rhyme was much more real to pioneer children:

Little Miss Muffet	*Along came a spider*
Sat on a tuffet,	*And sat down beside her,*
Eating her curds and whey;	*And frightened Miss Muffet away.*

Courtesy Russell Foster

BURNS' BLACKSMITH SHOP, EAST OF COBOURG, 1910

Courtesy Mrs. George Irwin

CHARLES WHITE'S BARBER SHOP, GORE BAY, 1890's

In the compartment at the rear was the village's first bathtub, available to customers at 25 cents.

WHILE the earliest settlers had to undertake many of the crafts and processes described in this book, it was not long before men began to specialize in trades in the towns and villages. Philemon Wright, founder of the city of Hull, was responsible for the introduction of many workshops and manufactories in his settlement. He writes as follows of the changes in 1804:

This year I commenced building a blacksmith's shop, which is large enough for four workmen to work in (it contains four pairs of bellows, which are worked by water, also four forges); likewise a shoemaker's shop and a tailor's shop, with a large bakehouse; all those establishments give employment to a great number of workmen. Before I established these different branches I was obliged to go to Montreal for every little article in iron work, or other things which I stood in need of. . . . And I also commenced a tannery for tanning of leather upon a large scale.

Wright had also grist-mills, sawmills, a distillery, a hemp mill, and elaborate barns in his settlement, which was probably the most complete and self-sufficient of all 'back township' settlements. The towns along the front had meanwhile large numbers of tradesmen, though at first the unskilled were often to be found imposing on their fellow-citizens. One settler wrote: 'Mason work is dear, and like the others, every fellow who knows the shape of a trowel or the handle of a stone hammer from the head dignifies himself into a mason, and fearlessly undertakes a chimney which often falls before half finished.'

But competition soon forced unskilled workmen out of business. A census of the town of Cobourg in 1857, when it contained between 6,000 and 7,000 inhabitants, listed five contractors, sixty-six carpenters and builders, and seventeen cabinet-makers, while there were also numerous wagon-

Public Archives of Canada   Lieut. Philip Bainbrigge

COBOURG IN 1840

Upper Canada Academy, later Victoria College, may be seen
in the rear at the left.

makers, carriage-makers, and shipwrights. Many tanners, tinsmiths, plasterers, and painters carried on their work in the
town, and the trade of blacksmith, now a curiosity, had
twenty-two representatives. While brick buildings had only
begun to be common, there were nine bricklayers in Cobourg.
as well as five masons and two moulders. Shoemakers numbered twenty-two and tailors twenty-four, but there was only
one clothes-cleaner, one hatter, and one dyer. There were
three saddlers and harnessmakers, and nine coopers, whose
business it was to make barrels, pails, churns, and vats, using
staves and hoops and doing most of the work by hand.

Distillers and brewers numbered four in Cobourg, while
several foundrymen and machinists were employed in local
manufactories. Two marble factories, presumable largely engaged in making tombstones, two chandlers and soap-makers,
and one rope-maker were listed among the manufacturers;
while two gas-fitters and a number of engineers, switchmen,
and car inspectors indicate that a gas company and railways
had been inaugurated. In addition to all these, there were
among the numerous merchants of the town many craftsmen
like bakers, confectioners, jewellers, watchmakers, dress-
makers, and milliners, and the modern photographer was re-
presented by one 'daguerrean artist', whose pictorial products
included daguerrotypes, ambrotypes, and crystaltypes, which,

Paris Public Library   Paul Wickson

THE JACOB AHRENS POTTERY, NITH RIVER, PARIS, ONTARIO, 1883

Wickson, a pupil of Sir E. J. Poynter, painted the pottery a few weeks
before it was destroyed by a flood.

Courtesy Jeanne Minhinnick

GEORGE GILLARD'S BLACKSMITH SHOP

DAM AND FLUME, McGOWAN'S MILL, SAUGEEN RIVER

TANNERY IN OMEMEE, 1890's

it was proudly announced, could usually be taken in cloudy as well as fine weather.

The small shop where one or a few tradesmen produced and sold their product lasted longer in some trades than in others, and we have still many independent carpenters, painters, plasterers, and other craftsmen. But there was often a change in status to skilled labour receiving wages from an employer, frequently in a factory where machines replaced hand work. There were soon axe factories and pail factories, chandlers and soap-makers, furnaces (forges) and foundries, sawmills and planing mills, plaster mills and flour mills. Large woollen mills superseded the smaller, though the carding mill long survived to prepare wool for farmers who continued to spin their own yarn, and the fulling mill to work up the cloth. There were oat mills and barley mills, distilleries and breweries, asheries and tanneries, flax works and cloth factories. There were many potteries where stoneware and earthenware were made, and occasional establishments where porcelain-gilding and enameling were carried on.

As the years passed, most of these local establishments went out of existence when new methods and new products replaced the old. Modern furniture factories, concentrated in a few cities, have replaced almost all of the cabinet-makers of seventy-five years ago. A few large firms make our supply of soap, paint, and many other products. We have still shoe repairers, but few of them now make shoes; similarly most tailors have become agents for large clothing-makers. A few large textile factories have replaced the many small firms of an earlier day; others, like potasheries and candle factories, have been entirely or almost entirely displaced. In a word, the day of the small tradesman working by hand in his own shop has largely given way before the onslaught of factory and machine.

With these changes have come others affecting personal characteristics. The sturdy independence of the pioneer who supplied most of his own needs has given way to a more complicated way of life in which no community lives unto itself. But if we are as a result less self-reliant than our ancestors, we enjoy the compensation of innumerable processes, products, and services that would have amazed our pioneer forefathers.